Fractured Ecologies

Fractured Ecologies

CHAD WEIDNER

editor

EyeCorner Press

ACKNOWLEDGMENTS

The construction of *Fractured Ecologies* was a daunting, long, and sometimes sporadic process. The assembly of such a heterogeneous and hybrid book was incredibly challenging. I had to balance personal allegiances, professional responsibilities, what at times seemed like thousands of emails, some diplomatic and some frank, the needs of hungry authors eager for feedback and exposure, and tragedies large and small. It was a long process. I would like to express gratitude to a number of people who helped get this book in print.

First, I thank Camelia Elias and the motley crew at Eye-Corner Press. Camelia's frank guidance, boundless encouragement, and incredible patience cannot be exaggerated.

My professional environment was conducive to the production of this book. University College Roosevelt provided funding for me to attend conferences, where I was able to test my ideas in front of critical-thinking colleagues, and for the associated copy-editing costs for this collection.

Thanks go to Joyce Bond for her sharp eyes and truly mad skills.

I am grateful to Utrecht University, which funded a post-doctoral position that helped me grow intellectually, and to Rosi Braidotti, especially, who really stretched my limits. Working with such a brilliant thinker was truly fantastic.

I'd like to acknowledge my inspiring visits over the years to the British Library, the University of Colorado, and the Small Press and Chapbook Collection at Naropa University's Allen Ginsberg Library, which introduced me to new classes of experimental literatures of resistance.

My thanks also go to Franca Bellarsi, with whom I presented a paper related to this project at the "Dwellings of Enchantment" conference at the Université de Perpignan, France, in June 2016.

Finally, I want to thank those people who continue to inspire me in ways they cannot fully appreciate. This book exists because you exist.

Table of Contents

Preface

Last summer, I encountered a bear in the northern part of Yellowstone National Park, and it got me thinking. I regularly visit the Rocky Mountains for camping with the kids, and bear sightings are not uncommon. Once I saw a grizzly bear absolutely shred an elk carcass. She tugged at the flesh and pulled it apart as if it were string cheese. Most bear sightings are far more ordinary: a mellow juvenile grizzly chewing on berries in the morning sun, a black bear in the far distance dozing in the afternoon shade, or a momentary brown flash amongst the trees. I've seen dozens of bears over the years, but last summer was different. After a hike, I navigated back to the rental and noticed cars were backed up into the main road, blocking traffic in all directions. Yellowstone can be a very busy place, and the infrastructure was not built for modern traffic. People were pointing up in my direction—pointing at me, I thought. Maybe they wanted my parking spot. However, this was not a case of parking envy. This was a legitimate bear jam. To my immediate right, perhaps only four meters away, a black bear paced along the ridge parallel to me. She grunted, and for one terrible and terrific moment, she turned and studied me with black penetrating nonhuman eyes. My ice-age DNA vibrated with recognition of the danger. I followed established guidelines to stay safe: talk in a calm human voice, slowly back away, and as my revered uncle Jim in Colorado always says, "Just make yourself look big." Thankfully, the bear was completely indifferent to my presence, and she soon vanished into the pasture ahead.

My near-bear experience got me thinking hard about nature, and the nature of my profession. I think the academic response to environmental problems is largely ineffective. So I wonder whether the intellectual power of the environmental humanities can bring the message to the people through experimental and environmentally sensitive poetry.

The idea of *Fractured Ecologies* is to participate in environmental praxis through literary practice. How does radical experimental writing contribute to the ways we think about ecology? This collection of papers, bent essays, and playful poetic impressions positions marginal aesthetic forms front and center. *Fractured Ecologies* is based on the premise that the potential of fringe texts remains overlooked in environmental discussions. Things are slowly shifting. Jonathan Skinner's *Ecopoetics* was a pioneering effort in the field, but unfortunately such interventions are uncommon. Sometimes efforts attempt to mimic the discourse of the academy, which limits reception. Wild avant-garde writing is a limited case of sorts, and the challenges in engaging with such forms are impossible to avoid. The idea of *Fractured Ecologies* is that rigorous and irreverent papers addressing experimental writing and other borderline manifestations in an environmental context are infinitely interesting and always fresh.

In 2013, I secured a contract from EyeCorner Press for this book and circulated a public call for papers. The interest was instant, global, and unexpected. *Fractured Ecologies* has been in utero since, and it is the result of years of rigorous thinking. In 2014, I co-convened the "Flow and Fracture from North America to Europe and Beyond" poetry laboratory at the Université libre de Bruxelles, Belgium, with Franca Bellarsi. It was an intense and inspiring event with imaginative literary thinkers and poets from many countries. In his electric keynote, Jonathan Skinner implored us to open our ears to the sounds around us, really open our ears. He then proclaimed, "Poets can move stones with song." This collection was assembled in that spirit.

Chad Weidner
Middelburg, the Netherlands
17 April 2020

Introduction: Deliberate Poetry as Ecological Survival Technique

CHAD WEIDNER

> *What is our purpose in this dark age?*
> *—Anne Waldman*

In recent years, ecocriticism has worked to help revive the humanities within the academy, but fractures remain. The publication of Jonathan Skinner's groundbreaking *Ecopoetics* Journal exposed the tensions between academic and poetic conceptions of literature as witness to the ecological crisis. *Ecopoetics* was new since it explored the creative/critical spaces between writing and the natural world with the acknowledgment of anthropogenic climate change. The phenomenon of climate change has long served as a source of inspiration for writers, and the critical literature widely recognizes this fact. But important questions remain. Can the humanities contribute to current debates surrounding climate change? Can creative activities help reconcile the many aesthetic and ethical questions that climate change forces us to ask? *Ecopoetics* journal participated in the wider shift in the humanities away from poststructuralist preoccupations and toward a return of sorts to scholarly and aesthetic engagement with nature and the environment, with the additional contemporary understanding that human activities threaten global ecological systems. Against this background, *Fractured Ecologies* explores the ways in which experimental poetry and prose contribute to how we can think about the environment.

Welcome developments in the ecological humanities have unlocked opportunities to work on serious social questions, but the humanities remain vulnerable. I see ecocriticism as the "study of the relationship between cultural texts and the environment conducted in a spirit of commitment to environmentalist praxis" (Weidner, *Green* 2). Simply put, ecocriticism is part of the wider movement toward the interdisciplinary environmental humanities. Ecocriticism covers an array of scholarly approaches that emphasize texts of all sorts in the context of the Anthropocene, climate change, and the Sixth Mass Extinction. Such a move was necessary. In a time of real crisis in the academic humanities, and there absolutely is a crisis, this is a basic question of survival. In 2018, Justin Stover declared, "The humanities are not just dying—they are almost dead." (See also Donoghue and Hayot.) We've got our backs against the wall. The alarming effects of climate change may force us to even question the value of literary studies altogether, and although the humanities are responding to the problem of climate change, more work needs done. Scholars working in ecocriticism need to examine entire bodies of cultural texts, and given the need for all academic disciplines to respond to the pressing challenges of climate change, their involvement is urgent. The future trajectory of ecocriticism is the subject of healthy discussion and passionate debate, and it is clear that it remains a vibrant, dynamic, ever-changing, and socially relevant subfield of humanities research. Although I remain fully committed to teaching and to the promises of my discipline, the grinding down of the academic humanities is admittedly frustrating.

More concerning is the broader intellectual response to increasing environmental problems, which remains far too slow, rigid, and even patronizing to environmental activists on the ground who are willing to get their hands dirty. Given the real vulnerability of the academic humanities today and the specter of unimaginable environmental crises that exceed the human capacity to fully fathom, the stakes are incredibly high, and the inability to connect to a wider public discourse remains a serious problem. Indeed, the academy fails to communicate the urgency of the ecological crisis to the public imagination because of both an overly orthodox academic climate and the wider prioritization of economic growth. Swedish teen environmental activist Greta Thunberg has articulated the need to address the ecological crisis to the public mind far better than the entire body of ecocriticism to date. This fact is both astonishing and disheartening. And while I fully recognize that ecocriticism has made an impact within the academy, at the same time it has been relatively ineffective in connecting to the wider public debate. Yet the inability of the sciences to do so is alarming for other reasons. The humanities are attempting to cross disciplinary boundaries in addressing matters of great urgency. The natural and hard sciences seem to plod along with increased urgency but also in seeming insularity. The situation is not entirely the scientists' fault, since corporate investment in climate crisis deniability interferes with the ability to carry out impartial and independent research, as well as to form and communicate a coherent message. I have no doubt that free-market think tanks, mercenary scientists, and corporate endowed professorships obscure the real issues from the public. Current climate denial strategies are based in part on misleading arsenic, asbestos, and tobacco tactics from the past.

Fig. 1. "Dr. James P. Campbell's Safe Arsenic Complexion Wafers," ca 1890, from National Museum of American History

Global corporate entities have access to incredible amounts of consumer data and have developed robust and sophisticated social media strategies to bend the public discourse to théir interests. Governments, too, are involved in shaping public discussion, but imagine what the PR offices at Gazprom and Shell are up to. In 2018, the Intergovernmental Panel on Climate Change confirmed that climate change is real, mapped out different disturbing potential scenarios, and recommended robust initiatives going forward (Allen et al.). But the scientific consensus is often overlooked by the public or drowned out by corporate distortion and political white noise. The simple fact is that both the humanities and the sciences have incredible difficulty communicating to the public about the issue of climate change. This is where creative practices can respond to the problems of climate change in new ways.

Fractured Ecologies thus engages with contemporary social concerns about climate change and on wider issues of accountability and civic responsibility. Through critical and creative reflection, this collection of papers and literary fragments seeks to negotiate the uncertainty that climate change generates by demonstrating ways human culture responds to social problems. I wonder why so many clever people work so much harder on coining ephemeral neologisms than they do on saving the world.[1] I wonder whether creative-thinking practices can help stir a fatigued public from "the narcotic tobacco haze of Capitalism" (Ginsberg, *Howl* 13). I wonder whether creative endeavors can translate urgent ecological considerations to more people. So what *Fractured Ecologies* attempts to do is interrupt the academic discourse, or at least question common assumptions (see also the disruptive strategies of the Extinction Rebellion). Such a change is decades in the making. Stover says humanists no longer have to justify our existence, suggesting instead, "All we need to do is put our hand to the plow." It's time to roll up our sleeves. It's time to get our hands dirty.

The green humanities make an impact, while the wider academy does not fully recognize the full potential for humanistic interventions into ecological questions, though history shows other paths are possible. Environmental humanists are genuinely concerned with the state of our planet. This much is clear. They are rightly troubled by corporate participation in the present circumstances and are disturbed by distressing amounts of political ignorance.[2] Scholars are troubled by corporate corruption in democratic systems overriding utilitarian human interests. Furthermore, generally speaking, the academy remains rather skeptical about what exactly humanists can bring to the table on the issue of environmental change.

> While there certainly seems to be a greater focus on sustainability within institutions, which is a welcome step, many of these developments are driven by a purely managerial and natural sciences perspective without incorporating [the environmental humanities] to any comprehensive degree. Therefore, the current situation requires a shift of sorts towards the incorporation of the all-important human dimension of environmental thought. (Weidner et al. 20)

Even more concerning is the ever-expanding and uncontested growth of hypercapitalism to produce more stuff without considering the very real material limits that now confront us. Greta Thunberg is spot on when she laments, "All you can talk about is money and fairy tales of endless economic growth" (par. 1). So a much more radical shift must occur if we're going to map our way out of the current predicament. The strangulation of the humanities under the guise of austerity and greater efficiency are not the only problems within the academy. In an incredibly competitive academic culture, scholars are sometimes willing to sell themselves out. This results in endless and predictable infighting, when amplified levels of cooperation are needed now more than ever.

Many professors claim to be champions of the common people and nature yet fail to consider the ways in which their meat-heavy diets, frequent plane trips, takeout coffees, and slave-labor laptops reinforce the very systems they claim to resist. Many are urban creatures who lack practical knowledge of the natural world: the antiseptic and healing properties of tree sap, the zest of morning acorn coffee, the olfactory overload of millions of lodgepole pines, the echo of wolfpack howl, the temporal slippage that the wilderness can generate in humans, or even the faint funk of big-game scat. Environmental scholars need to take nature seriously, and as more than just a theoretical problem to debate in the academic game of one-upmanship (see Robisch, Garrard, and Gaard).[3] The transcendentalists showed that what may seem like opposing ideological vantage points can work toward new forms of intellectual-nature convergence.

Eurocentric thinking is partly to blame for the current circumstances, but it is simply no longer acceptable for academics to remain neutral on environmental issues. I served on the scientific committee for the first ecocritical conference held in the Netherlands in 2010. Scholars at the event were quite critical of American ecocritics for not engaging enough in theory. Fair enough; different strokes and all that. However, I also met a Spanish professor who explained to me, quite sincerely, that he had no need to engage with activist thinking as American ecocritics do, since he was a "pure intellectual." What a bunch of baloney. I wondered if the guy even cared about the ecological predicament *at all.*

Louise Westling reasons that true "ecological humanism" can "restore appropriate humility" (qtd. in Arnold et al. 1104). Such a view is especially important in thinking about the theory-praxis divide as it exists within ecocriticism today. Researchers need to write clearly to connect to a wide audience, stay modest, and cooperate with anyone willing. Other eco-thinkers are more critical. Gary Snyder asserts that Eurocentric logic forces us to think of nature as a commodity and protests that "deconstructionists who believe that nature is merely a 'social construction' are deluding themselves." He further rebukes "high-paid intellectual types," who participate in what he calls "moral and political shallowness" (32-33). Kip Robisch goes further: "We're being poked with pens by people who have such a limited curiosity about the world beyond their own minds that they've greenwashed ecocriticism" (703). These are harsh criticisms, indeed, though they are not entirely unjustified. The situation is so dire, Robisch says, that the time has finally come to "start monkeywrenching the theory machine" (700). The climate situation is now so urgent that the theory wars must remain secondary. The larger existential dangers today are just too great to waste energy debating abstract buzzwords and deriding potential allies. Thus remaining neutral on climate change is simply no longer a viable moral or intellectually defensible position.

Fractured Ecologies declares that the academy needs a bit of monkeywrenching, in the spirit of nonviolent environmental praxis. Anne Waldman asks an essential question and at the same time provides a possible response: "How do we reinvent resistance? Let us be worthy of our demise." Let me say up front that the monkeywrenching I envision is completely nonviolent but highly disruptive. An appropriate response, in my view, draws on the desperate artistic prayers of the historical avant-garde when confronted by their own existential industrial-age threats. One hundred years ago, an "experimental negative aesthetics arose to counter the age of industrial chemical warfare, and today the alarming and mounting dangers of environmental change again force us to rethink the form and function of literature" (Weidner, "Glorious" 1). *Fractured Ecologies* embraces the urgency of the Dada call to engage in wild aesthetic revolt in a desperate attempt to save ourselves from certain death. I see disruption of mainstream discourse as a legitimate form of civil disobedience and dissent in the radical environmental spirit of Abbey's *Monkey Wrench Gang* and the contemporary *Extinction Rebellion* (see also Spierling). This has been a long time coming.

Fractured Ecologies was conceived as an extension of my academic interests in the Beat Generation, and ecocriticism can still learn from the Beats and other avant-garde movements. This project grew from questions emerging during my research on ecocriticism and the Beat Generation. The Beat Generation was a transnational group of postwar literary bohemians who promoted a poetics of resistance. Some believe the Beats have been canonized, though I challenge anyone to find William Burroughs's and Allen Ginsberg's most provocative works on common university reading lists.[4]

Ecocritics could examine more aspects of the Beats, which as a group remain underappreciated in the field of ecocriticism, with exceptions. Gary Snyder, Jack Kerouac, and Michael McClure were the Beat writers most clearly interested in "reconnection with the natural world," though William Burroughs occasionally "demonstrated a yearning for a return to a more pastoral—and even, at times, more primitive—existence" (Phillips 125). So my specific research tries to deal with Burroughs in this context. However, more needs done.

Grace and Skerl argued as recently as 2018 for wider academic recognition of the Beat endeavor, and it would be fitting to bring current trends in ecocriticism to the group. Even the Beats themselves considered the ecological impulse to the movement an essential feature of the project. In "A Definition of the Beat Generation," Allen Ginsberg emphasizes their contribution to the "spread of ecological consciousness" and the idea of "a fresh planet" (52). So the origins of this project were rooted in my interests on William Burroughs and the Beat Generation, and during my research, I came across many corresponding aesthetic movements that can really benefit from environmental reflection.

The basic idea of *Fractured Ecologies* is fresh and suggests that experimental poetic forms reverberate with alternative ways of thinking about the world as compared with conventional literary forms. *Fractured Ecologies* is new for a number of reasons. It adds to the humanistic knowledge base, improves our wider understanding of culture by engaging unfamiliar forms, and challenges trends within ecocriticism by focusing primarily on experimental texts. Interesting is that this collection explores aesthetic forms that might on the surface show little obvious apparent environmental potential.

The orientation of the book is also comparative, diverse, and transcultural, thus positioning it on the front lines of the environmental humanities. Finally, this book challenges assumptions about the ways literature is built from the cultural material that surrounds us. Jed Rasula says wild avant-garde practices can contribute to wider forms of "biodegradable thinking" (43). If he is correct, then studying experimental forms offers news ways of contemplating the world we inhabit. This collection assumes that literary texts should not be studied in isolation, and that significant environmental discourse exists between texts and larger cultural and societal systems. The dialogue between text and culture thus clarifies larger macro processes about ways humans think about the environment.

This project asserts that in the age of ecological crisis that various modes of cultural involvement are mediated through representations and interpretations of literature. *Fractured*

Ecologies therefore sees literature as a ripe and targeted space for environmental dialogue. A literary scholar might ask how the themes expressed in specific literary texts echo concerns about climate change. Such questions need to be asked, and remain central to first wave ecocritical studies at present. However, this project goes further by expanding the potential range of ecocriticism far beyond conventional environmental texts.

Concepts that helped shape the idea of this book include ecopoetics, post-pastoralism, shallow ecology, guerrilla communication, glorious plagiarism, and comic dark ecology. For our purposes, ecopoetics can be seen as inquiry into "creative-critical edges between writing (with an emphasis on poetry) and ecology (the theory and praxis of deliberate earthlings)" (Skinner front matter). Post-pastoral texts (e.g., Gifford) remain aware of the complexity of the historical pastoral yet also recognize contemporary human involvement in destructive processes in the real world. Shallow ecology (e.g., Naess) suggests the environmental situation will improve if one makes superficial token environmental gestures. Shallow environmentalism thus addresses only the symptoms of environmental problems rather than the root causes. Guerrilla communication involves appropriation and adoption of official discourse, which is undermined, subverted, and fed back to the receiver using familiar language, but in a much more hostile, modified form (Tarlo). Guerrilla communication is particularly suited to exposing peculiar and previously unknown subtexts, as well as expressing the urgency of social problems.

Fig. 2. "E$$o" – a visual example of guerrilla communication; source Greenpeace, 2007

In the example in figure 2, each letter s has been modified into the dollar sign, $, which destabilizes and subverts the official logo of the oil company Esso. The new message conveyed is that the primary concern of the company is profit. Glorious plagiarism is an amplified form of homage through linguistic appropriation and rerepresentation. The difference between guerrilla communication and glorious plagiarism is the degree to which the source texts are venerated. The recycling of a source text and refeeding of the modified text to the reader can work to amplify the original message or emphasize entirely unknown subtexts by the appropriating agent through mediation. Comic dark ecology can be seen as a form of humor that works as an effective strategy in emphasizing the effects of real ecological ruin in the face of truly apocalyptic scenarios that are difficult for humans to fathom. This collection was conceived in part by thinking about these concepts, and the next step is to consider other fringe forms.

There are nearly limitless connections between the Beats and corresponding experimental movements. These include the British Poetry Revival, the Fluxus movement, the Dutch *Vijftigers* (the "Fiftiers"), and the Czechoslovak New Wave. There have been so many incredibly creative forms and movements that deserve further reflection that it can be a little bewildering. Think of aleatory writing and related practices that rely on chance procedures. For example, can tarot reveal new forms of human intuition? Can the exquisite corpse method force us to think in new ways? Altar poetry is not new, but the inventiveness of the practice deserves consideration, as do so many other experimental and unconventional forms: assemblage, ASCII computer art, bizarro fiction, collaborative processes including comic jam, the cut-up and fold-in, calligram, chance procedures, concrete poetry, Dada writing, dictionaraoke, digital poetry, the exquisite corpse and other parlor games, fax art, Fluxus poetry, found texts of all kinds, fragments and remnants, glorious plagiarism, graffiti and wildstyle, guerrilla semiotics, haptic poetry, imagism, L=A=N=G=U=A=G=E poetry, endless varieties of mechanical narrative agency, pictography, psychography, round-robin texts, runes, sound poetry, surrealist writing, visual poetry, words-in-freedom, and still others that to date have gone completely overlooked in contemporary environmental discourse.

Fractured Ecologies is organized into three overlapping sections: proto-conventional, bent, and ruptured. The first section of papers in the collection can be described as *proto-conven-*

tional, a label that is not meant to be pejorative in any sense. Papers and creative pieces here deal with aesthetic structures that deviate to a degree from but still share many characteristics with orthodox forms. Most readers will recognize papers in this section as similar to what they have seen elsewhere, though with some differences. An example proto-conventional text may be rather orthodox in structure but unconventional in approach or content. The more traditional academic papers in this collection are placed here. The second section of the book includes bent forms. Bent constructions are essentially hybrid forms that retain some conventional characteristics but are more investigational than what is typically found elsewhere. An example bent paper might play with the structure of an academic essay in new ways that would never be considered acceptable by mainstream academic publishers. These papers may or may not include a bibliography or a bibliography of sorts, but the form of the paper is unquestionably hybrid and violates conventions and expectations. Alternatively, creative forms that bend predictable structures are positioned in this part of the book. The final section of the collection is best described as *ruptured.* Writing forms here bear little resemblance to conventional poetic structures in either form or content.

Admittedly, there are no clear distinctions between the three different parts of the book, and the boundaries are liminal and overlap. Therefore, I made the conscious decision not to insert artificial boundary markers between sections. *Fractured Ecologies* thereby asks readers to participate in the construction of the book. Joshua Schuster suggests that a

"Conceptual Ecopoetics" involves "learning what we can do with concepts and their various junctures with materials and landscapes rather than trying to territorialize" poetic spaces (227). In other words, the three sections respond to the reader. As the papers engage in the larger conceptual questions, *Fractured Ecologies* plays with ideas of authorship and readership. Thus both contributors and readers investigate key issues of agency, authenticity, autonomy, and environmental praxis.

The hope is that *Fractured Ecologies* splits open a crack. It openly questions how experimental textual aesthetics and other borderline manifestations might engender a sense of environmental responsibility in the real world. The collection seeks to shed light onto the linguistic and literary crossing of boundaries that experimental aesthetics generate and the ways in which liminal aesthetic textual forms allow for potential cultural recovery. This is particularly interesting in investigating fringe forms that defy genre or generic classification, in the context of ecology. I acknowledge that the eccentric approach of this book might strike some as contentious, deviant, or even bizarre. That is the goal, since it will then generate much-needed debate about the practice of ecocritical and creative research in the future.

Endnotes

[1] The Anthropocene is an example of a hot concept that has gained traction in environmental discourse (Crutzen 23). Essentially, the idea is that humans created a new geological age with the start of the Industrial Revolution, when significant amounts of carbon were introduced into the atmosphere. However, the term Anthropocene is undeniably focused on human considerations. Donna Haraway suggests the more inclusive term "Chthulucene" (*Staying*). Other related concepts include the Capitalocene (Moore), Plantationocene (see Haraway, "Anthropocene"), Pyrocene (Pyne), and more recently, the Traumacene (Garza). Additional spin-offs are no doubt forthcoming, including the Coronacene.

[2] Shortly after impeachment for abuse of power and obstruction of Congress, Donald Trump shared his opinion on windmills at a reactionary Turning Point USA rally in Florida on December 22, 2019: "I never understood wind. You know, I know windmills very much. I've studied it better than anybody I know. It's very expensive. They're made in China and Germany mostly—very few made here, almost none. But they're manufactured tremendous—if you're into this—tremendous fumes. Gases are spewing into the atmosphere. You know we have a world, right? So the world is tiny compared to the universe. So tremendous, tremendous amount of fumes and everything. You talk about the carbon footprint—fumes are spewing into the air. Right? Spewing" (White House).

[3] I recognize Robisch's legitimate criticisms about theory-obsessive ecocriticism. His desperate plea resonates now more than ever. I also appreciate Estok's 2009 call for eco-activism, though I am left with doubts about his specifically pointing to vegetarianism when people in the West contribute to the ecological crisis in so many other ways. Gaard's appeal for more feminist ecocriticism gained momentum, and the critical literature has respond-

ed. However, the many quarrels within ecocriticism seem rather trivial while Australia is on fire (see also Wallace-Wells).

[4]   In fact, John Weir's "10 Most Hated Books" lists both Ginsberg's *Howl* and Burroughs's *Naked Lunch.*

Works Cited

Abbey, Edward. *The Monkey Wrench Gang.* Lippincott, 1975.

Allen, Myles R., et al. *Special Report: Global Warming of 1.5°C.* Intergovernmental Panel on Climate Change, 2018, www.ipcc.ch/report/sr15/.

Arnold, Jean, et al. "Forum on Literatures of the Environment." *PMLA,* vol. 114, no. 5, 1999, pp. 1089-104.

Crutzen, Paul J. "Geology of Mankind." *Nature,* vol. 415, no. 6867, 2002, p. 23, http://dx.doi.org/10.1038/415023a.

Donoghue, Frank. *The Last Professors: The Corporate University and the Fate of the Humanities.* Fordham UP, 2008.

Estok, Simon C. "Theorizing in a Space of Ambivalent Openness: Ecocriticism and Ecophobia." *Interdisciplinary Studies in Literature and Environment,* vol. 16, no. 2, 2009, pp. 203-25, doi:10.1093/isle/isp010.

Extinction Rebellion. *This Is Not a Drill: The Extinction Rebellion Handbook.* Penguin Books, 2019.

Gaard, Greta. "New Directions for Ecofeminism: Toward a More Feminist Ecocriticism." *ISLE,* vol. 17, no. 4, 2010, pp. 643-65, doi:10.1093/isle/isq108.

Garrard, Greg. "Ecocriticism." *The Year's Work in Critical and Cultural Theory,* vol. 19, no. 1, 2011, pp. 46-82, doi:10.1093/ywcct/mbr003.

Garza, Eric. "Awakening to the Traumacene." *Age of Awareness,* 29 May 2019. https://medium.com/age-of-awareness/awakening-to-the-traumacene-8d5dcb92ea7f. Accessed 15 Jan. 2020.

Gifford, Terry. *Pastoral*. Routledge, 1999.

Ginsberg, Allen. "A Definition of the Beat Generation." *Friction*, vol. 1, no. 2-3, 1982, pp. 50-52.

———. *Howl and Other Poems*. City Lights, 1956.

Grace, Nancy M., and Jennie Skerl. "Standing at a Juncture of Planes." Afterword. *Hip Sublime: Beat Writers and the Classical Tradition*, edited by Sheila Murnaghan and Ralph M. Rosen, Ohio State UP, 2018, pp. 271-76.

Greenpeace. "Exxon Still Funding Climate Change Deniers." 2007, https://wayback.archive-it.org/9650/20200406054534/http://p3-raw.greenpeace.org/international/en/news/features/exxon-still-funding-climate-ch/.

Haraway, Donna. "Anthropocene, Capitalocene, Plantationocene, Chthulucene: Making Kin." *Environmental Humanities*, vol. 6, no. 1, 2015, pp. 159-65, doi:10.1215/22011919-3615934.

———. *Staying with the Trouble: Making Kin in the Chthulucene*. Duke UP, 2016.

Hayot, Eric "The Humanities as We Know Them Are Doomed. Now What?" *Chronicle of Higher Education*, 1 July 2018.

Moore, Jason W. "The Capitalocene, Part I: On the Nature and Origins of Our Ecological Crisis." *Journal of Peasant Studies*, vol. 44, no. 3, 2017, pp. 594-630, doi:10.1080/03066150.2016.1235036.

Naess, Arne. "The Shallow and the Deep, Long-Range Ecology Movement: A Summary." *Inquiry*, vol. 16, no. 1-4, 1973, pp. 95-100, doi:10.1080/00201747308601682.

National Museum of American History. "Dr. James P. Campbell's Safe Arsenic Complexion Wafers." ca 1890, http://n2t.net/ark:/65665/ng49ca746ac-5ea5-704b-e053-15f76fa0b4fa.

Phillips, Rod. *"Forest Beatniks" and "Urban Thoreaus": Gary Snyder, Jack Kerouac, Lew Welch, and Michael McClure*. P. Lang, 2000. Modern American Literature, vol. 22.

Pyne, Stephen. "Big Fire; or, Introducing the Pyrocene." *Fire*, vol. 1, no. 1, 2018, p. 1, doi:10.3390/fire1010001.

Rasula, Jed. *This Compost: Ecological Imperatives in American Poetry.* U of Georgia P, 2002.

Robisch, S. K. "The Woodshed: A Response to 'Ecocriticism and Ecophobia.'" *Interdisciplinary Studies in Literature and Environment*, vol. 16, no. 4, 2009, pp. 697-708, doi:10.1093/isle/isp096.

Schuster, Joshua. "Reading the Environs: Toward a Conceptual Ecopoetics." *Ecopoetics: Essays in the Field*, edited by Angela Hume and Gillian Osborne, U of Iowa P, 2018, pp. 208-27.

Skinner, Jonathan, ed. *Ecopoetics.* 7 vols. *Periplum*, 2001-9.

Snyder, Gary. "Is Nature Real?" *Resurgence*, no. 190, 1998, pp. 32-33.

Spierling, Karen E. "The Humanities Must Go on the Offensive." *Chronicle of Higher Education*, 8 Dec. 2019.

Stover, Justin. "There Is No Case for the Humanities: And Deep Down We Know Our Justifications for It Are Hollow." *Chronicle of Higher Education*, 4 Mar. 2018.

Tarlo, Harriet. "Recycles: The Eco-Ethical Poetics of Found Text in Contemporary Poetry." *Journal of Ecocriticism*, vol. 1, no. 2, 2009, pp. 114-30.

Thunberg, Greta. "If World Leaders Choose to Fail Us, My Generation Will Never Forgive Them." *Guardian*, 23 Sept. 2019. https://www.theguardian.com/commentisfree/2019/sep/23/world-leaders-generation-climate-breakdown-greta-thunberg

Waldman, Anne. "The Beat Legacy in the Anthropocene." *European Beat Studies Network*, 29 Oct. 2015.

Wallace-Wells, David. "Global Apathy toward the Fires in Australia Is a Scary Portent for the Future." *New York: Intelligencer*, 31 Dec. 2019.

Weidner, Chad. "The Glorious Plagiarism, Trash Aesthetics, and Ecological Entropy of Cryptic Cut-Ups from *Minutes to Go*." *Humanities*, vol. 8, no. 2, 2019, p. 116, www.mdpi.com/2076-0787/8/2/116.

_____. *The Green Ghost: William Burroughs and the Ecological Mind.* Southern Illinois UP, 2016.

Weidner, Chad, et al. "The Emergent Environmental Humanities: Engineering the Social Imaginary." *Connotations,* vol. 28, 2019, pp. 1-25.

Weir, John. "The 10 Most Hated Books." *Advocate,* 24 June 1997, p. 91.

White House. "Remarks by President Trump at Turning Point USA Student Action Summit." 22 Dec. 2019. www.whitehouse.gov/briefings-statements/remarks-president-trump-turning-point-usa-student-action-summit-west-palm-beach-fl/?utm_source=link. Accessed 15 Jan. 2020.

'GOLD/ Is Recovered':
Maggie O'Sullivan and Environment

Harriet Tarlo

Maggie O'Sullivan is a radical poet, visual artist, publisher, and performance artist whose work first appeared in the late 1970s. She acknowledges a debt to the modernist tradition and, in her work, we can see and hear fragments, forms, and linguistic inventions reminiscent of Gertrude Stein, Ezra Pound, and Basil Bunting. Though her Irish inheritances are very significant, she can also be read within the Northern modernist tradition often neglected in the writing of the histories of poetry in the modernist tradition. I have argued elsewhere that Bunting's Northumbrian poetry is echoed in O'Sullivan's evocation (her calling forth) of the West Yorkshire moorland, where she has lived since 1988.[1] Influences aside, O'Sullivan is one of the most original poets in the UK today, having little care for fashions, aesthetic or ideological. She resists any one given vocabulary, poetic or otherwise, in favour of a unique and recognisable voice (or, more accurately, voicings) composed of multivalent use of language from numerous sources, often ancient or from 'the celtic fringe', sonic experimentation with minimal units of sound evoking the more-than-human world, and creation of puns, homonyms, and neologisms, often from recognisable fragments.[2] Her collaboration with the American poet Bruce Andrews, *EXCLA*, for instance, is derived from three thousand words, many 'found', handwritten on tiny pieces of paper, and exchanged between the two writers.[3]

O'Sullivan also works in mixed media and cites Joseph Beuys and Kurt Schwitters as influences.[4] From its beginning, her work has involved painting, drawing, photography, photo-copying, and incorporating found objects and materials into banners and collages, reproducing these on the covers and within the pages of her books. As with many of the best British experimentalists, her work was and is published in a diversity of small press publications. Thankfully, many of these ephemeral texts have been reissued in the Reality Street collection *Body of Work*, but it is important to look at the original versions to see how they operate visually.[5] As Marjorie Perloff and Mandy Bloomfield have noted, her books increasingly resemble artists' books, and this is particularly true of her Etruscan and Veer publications (Perloff 124; Bloomfield 11-12). Sound and performance are also critical to her work, chiming with her sensitivity to embodied world. O'Sullivan's dual emphasis on the visual and the oral has led to her work being read through the filter of concrete poetry, and indeed, she often performed, collaborated, and co-published with the important, prolific British concrete poet Bob Cobbing. The visual, the textual, and the performative come together in her publications and public readings. In performance, her relatively slight figure on stage contrasts with her strong voice, with its distinctive Northern cadence and hint of Irish roots.

It is a common misconception to regard experimental or avant-garde writing as elitist or isolationist. O'Sullivan's use of a Gertrude Stein epigraph to her poem 'Another Weather System' uncompromisingly demonstrates her commitment to 'the world':[6]

> And each of us in our own way are bound to express
> what the world in which we are living is doing. (*Shaman* 8)

We often find in her work that it is 'what the world... is doing' in multifarious interrelated ways that she exposes as much as 'expresses'. She has a powerful sense of contemporary inter-human atrocities, often referencing their historical origins, such as the potato famine and legacy of the Troubles in relation to contemporary politics.[7] She paid close attention to gender from very early works onward, with *Un-assuming Personas* (1985) featuring a critique/romance structured around a series of passages beginning with HER and HE. She was painfully aware of environmental issues and referenced, in particular, the treatment of animals from her earliest publications, as we shall see. Though she eschews simplistic polemical purposes for poetry, when asked in an interview with Dell Olsen about her affiliations with other poets, she specifically praises Cecilia Vicuña's 'eco-ethico politics of the earth'. As the interview suggests, her contribution is not through theoretical debate or polemic, but through embodiment within poetic text and performances. Thus she acts, as the best artists do, as a transformative conduit for her readers and listeners.

Although I shall focus on environment here, Maggie O'Sullivan's terminology, and her formal experimentations and the philosophy behind them, are the more powerful for not having an environmentalist agenda and not separating environment from other issues. As SueEllen Campbell points out, environmental literature is often in danger of embodying the narrowness it attempts to escape: 'It is increasingly clear to me that environmental literature in general... works partly by shutting out social and cultural complexities – an omission that's probably one source of the desire they embody and evoke' (24). Looking through an ecopoetic lens, I hope to tease out O'Sullivan's profound and complex critique and consideration of what are, in effect, the philosophical conundrums of ecological thinking: the human/nonhuman binary, anthropomorphism, hierarchies, exploitation, agency, and how language might engage with 'nature', a term that is too prevalent in the culture to ignore but one that we should use with care.[8]

I am wary about being prescriptive as an ecocritic, a common failing in the first flush of all new 'political' criticisms. Rather, I am offering O'Sullivan's work as a gift here to those struggling with environmental concerns or criticism, rather than using her as an exemplar for a whole new series of prescriptions as to what 'environmentally sound' or 'responsive' writing should be doing. In fact, when we look back at the period of her practice I consider here (1985-2003), we find that it anticipated the original, developing, and current concerns of what has become known as the 'environmental humanities,' in particular the recent burgeoning of "New" or "Vital" Materialism, Agential Realism, and the fast-developing fields of Animal and Plant Studies. This 'materialist turn' attempts to take us out of anthropocentric thinking, to help us to 'to raise the status of the materiality of which we are composed' and to recognise ourselves as part of vital materiality, to discern more-than-human vitality (Bennett 12, 13-4). I shall demonstrate that we can be taken deeper into acts of radical imaginings through poesis, not least because of avant-garde creative art's ability to exist in contradictions.

I shall refer here to three book-works: *A Natural History in 3 Incomplete Parts* (1985), *In the House of the Shaman* (1993), and *Palace of Reptiles* (2003). All three books are tripartite, each with its own trajectory of transformation taking place over its three parts, which I shall touch on, but I would argue that each also reaches outside its own book-space to be decipherable as part of a larger ecoethical process of questing that occurs throughout the three books. In an interview with Charles Bernstein, O'Sullivan contemplates writing as a 'spiralling' process, a considering and reconsidering of 'area[s] of investigation' that 'declare [their] own materials' (np).

I shall follow my own spiral here, beginning with a discussion of the long poem 'Doubtless', which closes the second part of *Palace of Reptiles*. I shall then look back over the previous two books and return to the first part of *Palace of Reptiles* in order to explore the materialities of text and image that bring us back to where we began.

The poet and the earth

A foundational question that artists have always asked (and sometimes answered) in their work is: What is the role of the artist in the world? This question has gathered intensity throughout the last few decades as it became entangled with our sense of mounting environmental crisis. 'Doubtless' emerged from 'a live art performance/collaboration of Poetry Sculpture Dance Sound and Movement' (72) and features three 'characters': the Painter, the Poet, and the Dancer. We could read these three as personae of sorts, offering insights into O'Sullivan's own creative journey and/or as exemplars of the different relations an artist might have to the more-than-human world[9]. The poem opens with the Painter (the only figure in the piece gendered as male), who, armed with 'Knife & Axe', seems to engage in an exploitative grasping and measuring of other beings and plants:

> He clenched with melt of closely, the Gorilla,
> again Birds, Dawns on his back, his arm reaching
> like this –
> a Sponge to Disrupt a Vast
> Weeping – a pod, a pollen
> Fetched handful,
> Rooted (32)

This stanza occurs in the midst of a series of stanzas focused on strong verbs: 'He moves to strap', 'he/ armours & treads', 'he measures and scrys' (30-33). This character is reminiscent of the <u>HE</u> of *Un-assuming Personas*, of whom one witty yet awful line reads, '<u>HE</u> shot & missed & called this act, <u>art</u>'.[10]

The Poet, following on from the Painter, is associated with the hare, the sacrificial animal, and gives birth to the text in a bloody sacrificial moment:

A TUMULT OF HOLDING –
SHE BELLY-SPLIT
<u>DON'T.NOT.NOT.DON'T.NOT.NOT.NO</u>—
TO THE WOUND'S DIAGONAL
DISPARATE HARMINGS – (35-36)

Through her identifications with wounded landscape, the Poet is 'BIRTHING THE WORDS I FELL THROUGH/ BEATING THE WAY' (34). When we come to the Dancer, described as 'In the East/ Opposite the Painter', the language shifts dramatically into a rhythmic, beautiful ease, without ugliness or awkwardness:

The Dancer –

Jink –
Jointed
Uprised & Birth –

Stretching, Strung

Plover, low, lowing in a far field
AY – PR – PRO – LONGED, LESQUING – OFF – OFF
(fell, fell, soft) (50)

Through her bodily movements, the Dancer is akin to animals in the poem, particularly the hare and fish, who dance
and weave throughout. The dancer fits *in* and *with* the more-
than-human world, dancing and singing, in air (the plover), in
earth (the hare), and in water (the fish). The fish-dancing passage of 'Doubtless' is one of the talismanic and celebratory
moments of O'Sullivan's poetry:

(FISH
DALLY
IN THESE GREEN DEPTHS –

THEY SHEEN & CROSS
& WHISPER
FORRESTS
OF ANIMAL
AT EACH CURVE IN THE WATER
THERE IS A CREATURE
THERE IS A DANCE
OF CREATURES
IN DANCE OF
WANDERING
DANCING
INTO
THE SUN – (41)

Here the human being can be, seems to be, part of the
dance of the living world: this, O'Sullivan attests, *is possible*.

In her suggestive essay, 'Earth, World, Text: On the (Im)
possibility of Ecopoesis', Kate Rigby argues against Jonathan
Bate's bold Heideggerian statement in *The Song of the Earth*

that 'things need us so that they can be named'. Rigby suggests that 'rather, it is we who need to name things so we can share understandings about what we perceive and value, what we fear and desire, how we should live and how we should die'. She goes further, however, in arguing that we should be using artistic expression 'to join in the exuberant singing, dancing, shape-changing, many-hued self-disclosure of phusis... We need poets not so much to draw things into Being through their song, but rather to draw us forth into the polyphonic song of our nonhuman earth others' (434). Like O'Sullivan's dancer, it is within this possible/impossible nexus of ecopoesis that I wish to continue the argument throughout this essay, looking for how O'Sullivan opens possibilities whilst acknowledging impossibilities.

Here, then, are three possible models for relating to the world around us for the purposes of creation: the sometimes exploitative measurer who brings his own tools to bear on the earth, one who identifies with the earth's pain and attempts to voice it (akin perhaps to animal liberationists who argue for sentience as the criterion for protection or rights of animals), and one who attempts to merge with the earth and engage with its nonhuman inhabitants. All these are present in the multitude of texts that explore nature', environment, place, and landscape. We can read them, then, as coexistent trends in art and poetry, as they are in the wider culture.

In reading them as three-in-one, we can also read 'Doubtless' as a journey, a rite of passage or ritual, in which the initiate is changed by or reborn through their experience of animistic identification, words, music, and dance. Thus the poem is testament to transformative change. The Painter begins the poem in a dominant and uncompromising position, 'The painter stands', but at the close of the poem, we find him:

> On his back –
> crevices, caves, ear's flow –
> small processions –
> (look-on & visit) – (56)

Here he is attentive: he hears and sees the detail, even the (to us) minuscule insect world, and although he does not grasp everything he observes, he looks on. He is a visitor, and a newly respectful one.

We might even argue that by the end of 'Doubtless,' the artist initiate has learnt their place in the world. I use this neutral pronoun because there is fluidity and fusion between genders in this reading of the poem. If the male Painter is at first associated with what might be seen as stereotypical aggressive and epistemological approaches and the female with what might be construed as an early eco-feminist, even essentialist, merging with 'Mother Nature' , then both must comprehend these gender traps, move on, and merge elements of their behaviours to find new ways to attend and act. The poem then offers a complex, ecoethical exploration of 'the' artist's role(s), but it is also alert to the difficulties and pain of transformation through earth encounters: after all, how many of us can be three-in-one or can journey from one to three and are willing to devote time, energy, and courage to work through all inherited perceptions in order to do so?

The earth of language

As is true of many transgressive poets, Maggie O'Sullivan locates her ethics and politics in the form-language nexus. When Scott Thurston asks her about her contribution to Irish politics, she replies, 'For me I couldn't possibly do it in a linguistically transparent way'. Later, in response to Thurston's question, 'Do you feel that your writing... has a political import by nature of its approach to language?' she responds more warmly: 'Absolutely... What all of us are doing in using radically imaginative language practices and procedures is very political and subversive' (9). We have come a long way from the 1990s, when some ecocritical thinking drifted toward valorising poetry that came as close as possible to nonfictional nature writing or persuasive ecological writing.

When Lawrence Buell, in his admittedly groundbreaking *The Environmental Imagination* (1995), argued that 'linguistically transparent' language was the best linguistic form to engage with environment and began the process of defining what he described as 'environmentally oriented' writing (168 and throughout), he inaugurated a phase of somewhat prescriptive and selective literary ecocriticism. This was influential on a trend in poetry criticism that followed early in the early 2000s, when critics such as Leonard Scigaj and J. Scott Bryson began to define and critique (or perhaps commodify) a new subgenre, rapidly becoming known as ecopoetry, and similarly attached to transparency. These views are rather old hat now, and the argument against them, and in favour of reading more widely in the more feral forms of the genre, has been made by critics including Dominic Head, George Hart, Kate Rigby, Ursula Heise, Richard Kerridge, Lynn Keller and Linda Russo. Jonathan Skinner, poet and editor of the journal *ecopoetics,* is a particularly important figure in this development.

Writers, readers, and critics of experimental poetry can still be frustrated by even the most subtle and knowledgeable of ecocritics. Greg Garrard, responding to Rigby's essay cited above, writes of 'the exquisite combination of Heideggerian astonishment at Being, postmodern ironic self-consciousness, avant-garde poetics and physical engagement' that she asserts is *possible* within certain poetries, examples of which, he argues, are hard to find (267). Maggie O'Sullivan is one he should surely read, alongside many other contemporary poets who do such work, Rachel Blau Duplessis, Frances Presley, Juliana Spahr, Carol Watts and indeed Skinner and Burnett who are also cited here as critics, to name a few.

Rigby's accounts of what such writing could look like might have been written about O'Sullivan, as her own poetics attest. In 'riverrunning (realisations' O'Sullivan draws together key linguistic motifs from her wider practice:

> The works I make Celebrate ORigins/ENtrances – the
> Materiality of Language: its actual contractions &
> expansions, potentialities, prolongments, assemblages –
> the acoustic, visual, oral & sculptural qualities
> within the physical: intervals between; in and beside.
> Also, the jubilant seep   In So of Spirit – Entanglement
> with vegetations, thronged weathers, puppy-web we agreed
> animals.   Articulations of the Earth of Language   that is
> Minglement, Caesura, Illumination. (*Palace* 64)[11]

This passage moves seamlessly from linguistic experimentation to the poet's intense, even eroticised, engagement
(entanglement, minglement) with vegetation, weather, birds
and animals. The phrase 'Articulations of the Earth of Language' brings together language and earth through voice,
articulation, evocation (literally, calling forth). It reverses the
expression we might expect, 'the Language of Earth', and puts
earth first. Immediately the phrase explodes with multiple
meanings, as is common in O'Sullivan's work. Earth itself is the
whole earth, the planet, but also earth as in soil. If language
is 'of earth', then, it is earthy, visceral, as O'Sullivan's surely is,
but it is also representative in some way of the earth through
its interactions with it. The American poet Marcella Durand
claims that poetry is unique in its ability, 'perhaps even obliga-
tion, to interact with events, objects, matter, reality, in a way
that animates and alters its own medium' (62).

Elizabeth-Jane Burnett has usefully applied Sehjae Chun's
reading of John Clare's bird poems to her own reading of
O'Sullivan, thus teasing out some similarities in the two poets'
'unique use of grammar, spelling and punctuation' as part of
a 'deliberate effort to free the human-centred imagination to
embrace a merged subjectivity' in which the poet-observer
and observed are brought together (Chun 56, Burnett 38). In
'Articulations of the Earth of Language,' O'Sullivan plays with
the idea that language can be a wild-grown product of the
earth in some way, possibly through its continual and chaotic
process of transformation, a notion that many 'nature writers',
from Henry David Thoreau to Gary Snyder, have played with.
As poet-critic Peter Middleton observes, all words are in fact

neologistic and hence 'hairy, fleshy, wild', only 'they usually dress to hide it'. He notes that this is also true of species, which are so often variants or intermediates, seldom 'pure' (114).

If any language has the potential to strip back and simultaneously expose some of its acculturated layers, then surely it is O'Sullivan's language that comes closest to this. In the reading of a seemingly simple nugget of language from O'Sullivan's work above, we find many meanings, often reversed, unexpected, and stimulating. As readers, we are in O'Sullivan's language as we are on the earth: it is never *possible* to comprehend it all, and it is up to us to become involved in its interpretation. This is itself one way in which her utterances can be said to be 'Articulations of the Earth of Language'. In performance, where we do not have the time to stop and analyse the flow of language, this is even more the case.

In 'riverrunning (realisations', O'Sullivan describes her move to rural Yorkshire in 1988 as follows: 'I stepped out, away from the city to the moorland impress of tongue' (*Palace* 67). This gloss on the 'Earth of Language' chimes with another Cecilia Vicuña phrase, 'To feel the earth as one's own skin', cited by O'Sullivan in her interview with Dell Olsen. The human or animal organ, the tongue, in becoming here the organ of the moorland also, indicates O'Sullivan's desire to sense, taste, and speak earth from a less limited human perspective, to be moulded by the moor, to live in it in the present tense, not describe or evoke it from the outside, thus moving beyond the binary distinction between human and nonhuman.

When asked by Olsen how she sees 'Voicing my body/ bodying my voicings'... in relation to the possibility of an eco/

ethico politics', O'Sullivan replies with another citation from 'riverrunning (realisations':

> What 'Making' – 'Unmaking' is / a Mattering of Materials (mo-
> tivations & practise) – Living to live in that Learning – Uncer-
> tain, Uncurtained TonguescapeSUNG. SHUNTS. ARM WE. Living
> Earth Kinships on the
> vast-lunged shores of the Multiple Body imbued with
> wide-awake slumberings & cavortings. Constructions.
> Intuitions. Transmissions. Radiations. Thinking.
> ATTENDING. (Palace 65)

Here we see O'Sullivan, like the Painter at the end of 'Doubtless', as a writer in and of the world, 'Learning' from and 'ATTENDING' to her 'Materials' as part of her daily practice. As is common in her work, that 'now' of natural processes and the artist's attendance to them is evoked by repeated present participles. In the face of writing such as this, it is not *possible* to perceive the so-called 'nonhuman world' as passive, wheth-er as objectified or as resource or as victim. Anticipating the theories of the more imaginative ecocritics such as Donna Haraway, more-than-human elements of 'nature' move from a position as resource or thing to become a co-agent in the processes of the world and the production of knowledge.

'Tonguescape' brings body, writing, and performance to-gether, uniting tongues in speech. That speech or language is a live, energetic, and emotive force. It does not try to 'capture' what is not human in a mimetic net, but attempts to release its incomprehensible energies into language. It is language that must give. This 'concrete' or 'material' quality of O'Sullivan's work, words frequently employed in critical commentaries on her poetry, pushes the reader into the earthly world with a force seldom seen in more traditional forms of pastoral poetry or ecopoetry. Words, references, and neologisms that conjure up the more-than-human world are constant in her work, pushing at the inadequacies of language in this area, trying to make it do more and be more, even as it expresses frustration at the difficulty of this 'saying':

> in happy Yellow lies light
> not so much
>   leaves, it is & strapped,
> volume, this (towards the)
> saying of, pushing soil,
> (towards the) saying of,
>   (towards the)
> *(A Natural History)*

These lines reveal a drive to speak 'natural history', to 'the saying of', yet one that can never fully be articulated, is only ever '(towards the)', in parentheses. Yet the force in O'Sullivan's work can be felt by the reader or listener to be almost super- or hyper-real, beyond what we can see, pushing us down into the earth 'under the yellow iris' *(Unofficial)*.

Language, the treasury

In *A Natural History in 3 Incomplete Parts,* a text developed through the 1980s, O'Sullivan poses many of the questions that she will spiral around for many years.[12] She challenges anthropocentric epistemologies about 'nature' in terms of both how we think and feel about it, raising metaphysical issues regarding 'nature' as a philosophical term through which humanity defines its difference and superiority as a subject in relation to a world of objects or even nature itself as an object. In so doing, she prefigures ecocritical debates conducted by Kate Soper in her influential book published a decade later, *What Is Nature?* (1995). As the title suggests, the first premise of *A Natural History in 3 Incomplete Parts* is that we cannot know our natural history. Bunting's line from *Briggflatts,* 'Follow the clue patiently and you will understand nothing', echoed in my ears when I first read it (53).

The Magenta version of the book, with its bright, collaged pages, appears like a scrapbook rather than a reference book, with newspaper cuttings about human atrocities pasted across its pages. There is humility and humour in the division of *A Natural History* into three parts called 'Incomplete', 'More Incomplete', and 'Most Incomplete'. The assumption that nature can never wholly be known runs throughout O'Sullivan's oeuvre. It complements the idea that we might learn *from* nature. In the page headed 'introduction of sound:/ introduction of sight:/ introduction of texture:' (reproduced here), O'Sullivan demonstrates how an effort of cramming more and more words relating to insects onto the page does not increase the sum of what is 'known' about the natural world. Yet simultaneously the page does suggest the sheer extent of lives, sensory processes and progresses of insects the generic word repeated and hence defamiliarised. The text is further ruptured by the separation of each word by a full stop, emphasizing specificity, but also interrelation and equality by the very closeness of each word to the other.

The poet and writer, Elizabeth-Jane Burnett, whose own work is clearly influenced by O'Sullivan, describes the effect of such techniques as 'a burrowing into the words, each full stop drilling down upon a surface, denting, damaging, changing it... the moving 'into' process suggested by the word as signifier more deeply through the addition of this staggered punctuation' (39). While Burnett draws attention here to the linguistic innovations and effects, as the word "burrow" suggests, it is precisely this that leads us further into a more immersive relationship with the more-than-human world of critters usually considered somewhat less than a so-called 'charismatic species'.

As Nicky Marsh has suggested, the headings also suggest an interrelation or even a score, such as those produced for Writers Forum, and hence a performative element to this page (86). Who is seeing/speaking here? Perhaps it is the Painter 'on his back' at the end of 'Doubtless' observing the minutiae, yet the speaker is obviously a reader, too, of all those overwhelming 'natural histories' that claim to know.

introduction of sound:	introduction of sight:	introduction of texture:

```
Pollen.Primitive.Sting.Hatch.Firebrat.Tubular.Realm.Phose.Pest.
Ant.Anatomically.Furnace.Infernal.Warm.pigs.Pipes."Know.Feelers.
Noise.Green.Glanding.silk.or.Venom.Wasp.Antennae.Lock.Button.
Earth.The Next.Step.&.Seal.INSECT.INSECT.INSECT.INSECTS.Bumble.
Bees.Butterfly.Moths.Brine-fly.Wellheads.Weevil.Water.else.
Spring.Field.GrassHopper.Distant.they.as.Great.Air.to.find.Lo.
Locust.would.be.longer.Longer.Fly.Beetle.Mosquito.Suck.third.
Cylinder.Mine.zone.Certain.seeps.Air.peed.Pale.Green.Butter.
House.Wasp.Hornet.Honey.Horse.Pupate.Longhorn.Winged.poor.
starch."joined.feet".Anthro.Arise.Appendages.Bristletail.Or.
Organs.of.touch.proper.legs.BODY.SECTIONS.stride.bearing.
wingless.thorn.Nectar.wing.cannot.fold.wing.their.wings.walk.
&.thorn.A.climb.Grasscry.Mayflies.Dragon.Cock.Cicada.Beetle.
others.Fossilfly.Earwig.Seeping.White.pine.bit.of.tiny.tan.
mite.Sarcophagi.Banded.Fire.Stag.Horned.She.Twig-flight.Grub.
to.wing.form.Pale.&.ate.&.sea.Solid.sap.Tasting.a.Daisy.Dusted.
Green.Lines.loom.ordinary.honey.burrow.Goldenrod.Nips.its.prey.
Bits.Insidious.Prime-gut.Those.does.Diptera.Speck-suck.Arachnida.
tick.spy.spy.my.my.speck.legs.Daddy.Thousands.of.small.complete.
EYES.Steer.sun.Strike.eye.to.Doom.Mouth.pats.abundant.hair.to.
parts.bend.appendages.Fuse.ganglia.Nerve.calls.Mound.acuity.
Corneal.Red.looks.that.glow.Whip.tympanic.Stridulate.Sand.chirp.
Snare-drum.No.true.voice.Creak.Creak.Moon.lights.wing.rasp.
filly.vein.spur.Honeydew.drip.crop.clean.combs.Visited.oil.fun.
finicky.link.to.cab.nasturtium.bedings.of.the.hair.lift.ball.
fuzz.Oriental.the.white.those.in.air.praying.Mantis.maxillae.
coil.blood.mandibles.bite.sight.stylet.Heart.&.Poison.tubel.
Pyloric.fly.ord.(A).crop.complicated.mouth.houses.prismatic.
feathery.feelers.Sole.sole.on.wing.tiger-beetle.snout-meat.
Syphon-scent.half-zipper.does.the.work.cuticle.hook.&.eye.
halterers.More.flap.An.aphid's.birth.decade.painted.lately.
coming.habit.nymphal.moult.g.g.finally.wing.such.queen.quick.
blade.stem.cracks.ASSIGN.all.the.time.Spotted.Swarming.Faster.
gregarious.Winged.Monarch.into.four.fur.thin.velvety.tightly.
to.two.did.not.raw.could.Flow.Canker.Handmaid.Harlequin.Place.
to.hid.to.hide.Junkyard.of.A.slit.comb.unit.of.beebread.even.
pebble.porch.ovals.of.Thousand.Ovals.Leaf.Be.B.Begins.Build.
Another.Mix.Begin.a.new.Dauber.Vaselike.Wasp.Each.ineach.See.
Sealing.Pebbles.in.the.first.fist.Caterpillar.an.eggg.it.Thin.
cuticle.as.temporary.mate.Queen.Soldiers.Qoe.Worker.Primary.
Hind-Gut.Valve.valve.VALVE.Hatchet.into.heat.nest-fret.Gain.of.
carbon.White.ant.Hexagon.Saliva.Tool-using.WASP.Leaf-Roller.
Caddisfly.Cylindrical.Inch-brick.Ambrosia.Tap.excrement.Soft-
wood.bark.insects.into.jaw.into.Snap.dune.Many.times.Moth.larva.
teeming.deadly.partial.back.teeming.predaceous.Game.of.catch.
bee.flies.rob.flies.robby.Ladybird.Cliff.edge.abdomen.periodic.
sting.Black.&.yellow.unpalatable.caterpillars.Pillars.Exquisite.
cerise.perfect.sum.Resemblance.to.gaudily.birch.decor.as.Puss.
Moth.Milkweed.Four-Eyed.Bug.Shade.to.Discourage.ENEMIES.Spined.
Poisonous.Honey.Sparks.inDark.Four-Pointed.Shot.Pattern.Hurry-
ing.Only.Eye.Comb.Propolis.Stung.load.their.tongues.bee.wags.
Waxen.See.Sell.Clinging.to.Wing.Condition.Root-laced.mud.Dragon.
Damsel.Caddis.Nay.May.Mosquioto.sub.soaring.wrigglers.Skating.
whirligig.Backswimming.Waterboat.to.take.Silver.like.Breath.It.
has.to.Stray.line.tool.too.a.clutch.keep.duped.culex.wings.awhir.
awhile.Instinct.In.lay.for.it.Nest.Roof.Honey-Carnage.rend.a.
```

Fig. 1. Page from An Incomplete Natural History,
Writers Forum, 1984.

.

Throughout the book, O'Sullivan plays on the idea of knowledge, in particular in the form of definition and classification. The section titled 'LEAD VOCALS' consists of an incomplete alphabet of pseudo-definitions, lyrical and luxurious, but also faintly mocking in its references to our human uses of natural 'products' such as food, herbs, and essential oils:

> Lemon.Slices.mourning.
> LAVENDER.
>
> melissa.muddle.midline.Mandibular.Magenta.Maiden.
> macadamia.the.
> MOUNTAIN.ROSE.
>
> neroli.New.orders.of.destruction.Nori.New.herbs.&.NOwhere.
> NOTHING.

Incompleteness and juxtaposition make these 'definitions' into a serious/playful mockery of human knowledge and approach, one that continues in subsequent sections: 'COLOUR PLATES', consisting of a list of names with only spaces where the pictures should be; '<u>READING WRITING (a DOCUMENTARY)</u>'; and 'an inventory of impermanent address to no god.' There is always another way to see, to attend, than these human classifications and descriptions, and no one way can ever succeed. The title of her poem 'Another Weather System', from *In the House of the Shaman*, flouts another systemising of a 'natural'. Read now, the poem carries a prophetic irony as the climate change crisis mounts, challenging our ways of perceiving weather.

(Seasons

Are Not
Agreed
So Violently) (20)

In reading or hearing O'Sullivan, our brains become activated and energised by the multiple associations each particle
of language (individual words, phrases, neologisms, manipu-
lated, truncated or extended words, puns, compound words)
creates. In juxtaposing a vast range of discourses, including
here the languages of the natural sciences, poetry, myth,
economics, slang, common speech, and folklore, O'Sullivan
extends far beyond the denotative into the wild ways of the
connotative range. The word 'divisionary', for example, used
in *A Natural History in 3 Incomplete Parts,* suggests a vision, but
a vision divided, which can never be whole and perfect, perhaps a diversion, too, as well as division itself distorted into a
descriptive adverb.

An activated reader is a thinking, responsive reader, open
to the transformative in language, and thereby the challenge
of political and environmental orthodoxies. An epigrammatic
line from *Palace of Reptiles,* 'Treasury Futures,' occurs near the
end of the poem 'Birth Palette,' which refers to over twenty different species of animals. It suggests the treasure of legends
and fairy tales, with 'Futures' adding a play on the language of
the city or economics. Both suggest human values, but both
are subverted in the presence of the other. The implication
here is that wealth is in the diversity of plant and animal species, if their futures are intact and unthreatened. The same

subversion of notions of worth through unexpected juxtaposition occurs in the book title itself, *Palace of Reptiles.* It sends me back to the earlier poem 'Of Mutability', from *In the House of the Shaman,* where we find 'whortleberry kingdoms', a phrase subverting not only notions of worth, but also of scale, size, and perspective (39). All these examples place human concepts of worth (via economic evaluation or 'resources') in stark contrast with the 'wealth' of the more-than-human, posing a challenge to dominant capitalist and commercial attitudes to the world. These are tiny examples of how O'Sullivan's language creates conundrums, embodies contradictions, and thereby escapes, whilst acknowledging, conventional thought structures and current ideologies. To return to the theme of the possibilities and impossibilities of ecopoetics and ecoethics, here the *realpolitik* and the creative new possibility are coexistent in textual practice.

As we have seen, O'Sullivan's groupings of words from these diverse preexistent and invented vocabularies do not appear only in two- or three-word phrases, but also in lengthy structures, often for several unbroken lines in which they attain an intensity through the sheer buildup of syllabic sound and multiple associations. It is through this sonic accretion, sometimes close to listing, that O'Sullivan creates a sense of energy, movement, and process: above all, change. Transmutation is, of course, central to ecological processes but is often missing from more static 'nature poems' devoted to freezing the moment rather than releasing it on into the next moment and the next and the next. O'Sullivan's poetry celebrates and perhaps even envies transformation in the more-than-human world. One of the things humans might learn from 'ATTENDING' to the world outside themselves could be to accept our own subjection to change. In 'Of Mutability', we find an ABC of 'natural' processes:

> Snipe. Ashet abraiding   Bitters beak   Conduction
> crystal   a common lacerated thickly   Early Spring
> That Came   summering   copple blunts, clyst
> seedless Bomba dampling   traces of human bit/
> triply stilled   a bleeding means
> a Dock
> growing to
> Begin. (Shaman, 38)

Here the rhythm and sound of common, uncommon, and created words merge in the present tense, present participle abraiding, summering, dampling, bleeding, growing process of spring into summer. It is but one tiny example of how intricate alliterative sound patterns and multiple connotations amass in effect, in part because we cannot immediately stranglehold the language in a denotative grip. We see immediately Snipe, Spring, and Dock. We may not know the archaic words 'abraiding', as to start up or awake; 'copple', as conical hill; or 'clyst', as clear stream. We may surmise all sorts of sound and meaning from the neologistic pairings of 'Bitters beak' and 'Bomba dampling'. There's a little nod to the alphabet here, too, the seemingly simplest of linguistic systems (Ashet/ Bitters/Conduction/Dock), but this goes off the rails, rather like our naïve belief in our naming of the seasons, another system where we continue to believe that one will follow another. Herein, many ecocritics have argued, lies the heart of our problem with accepting climate change as a reality—we cannot believe the seasonal patterns of our lives will change. It seems as simple as ABC.

Throughout 'Of Mutability' there is celebration and richness, confirmed by the final assured and affirmative lines of the poem:

GOLD

Is Recovered. (40)

It is this sense of language in a 'state of becoming', constantly morphing and growing, that, Robert Sheppard has noted, allows O'Sullivan to 'sculpt instability within the linguistic sign'. He further argues that this 'linguistic proliferation is the nearest you'll get to the processes of nature' (*Far Language* 51). In a later essay on 'The Performing and the Performed', Sheppard builds on this idea, arguing that O'Sullivan's use of language is akin to the 'polysemantics of nature', the 'interchangeable yet changing natural environment'. Sheppard's analogy between the processes of nature and the processes of linguistic play in O'Sullivan's work pose a challenge to those who expect 'nature writing' to be realist or 'transparent'. I am not wholly convinced by what I read as his perception of O'Sullivan's language-play as a speeded-up, idiosyncratic version of what he calls 'real' or etymological changes in language, and hence as a parallel to ecological evolutionary shifts. However, O'Sullivan's writing clearly uses language in such a way that the syllables, words, and phrases reflect the evolving processes around us in their energised transmutation, but also become forces to make us feel and think about how we relate to environment. Whatever their assessment of it, readers and listeners are never able to be passive in response to her work.

Kinship with animals

Attendance is one thing. Even with the acknowledgment that it can only ever bring partial understanding, it still implies a degree of anthropocentric emphasis, an evident interest in the reaction of the human being (writer and reader) to the material, more-than-human world around them, and this is what dominates most nature or landscape poetry. O'Sullivan goes further, however, if we follow her recurring phrase, 'kinship with animals', through her work. The phrase 'Precious Animal Kinship' is found in *Unofficial Word,* and 'Kinship with Animals' is the title of the second part of *In the House of the Shaman.* In the extract of 'riverrunning (realisations' cited above, she refers to 'Living Earth Kinships on the vast-lunged Shores of the Multiple Body' (65). In exploring this, O'Sullivan again anticipated new developments in the far reaches of Animal Studies. In particular she prefigures Haraway's thinking in *Primate Visions* (1989), through *When Species Meet* (2013), to her most recent Staying with the Trouble: *Making Kin in the Cthulucene* (2017). In Haraway's earlier term 'companion species' and later usage of "kin" we see a movement beyond merely a notion of species to that of interspecies in which all are relational, co-dependent and agential, but also emotionally attached.

In her introduction to *When Species Meet* Haraway, like Soper, uses the image of the dance, writing about how 'species of all kinds, living and not, are consequent on a subject- and object-shaping dance of encounters' (4).

In elaborating O'Sullivan's 'kinship with animals' it is instructive to look at her answer to Dell Olsen's question about animals in her work:

> I spend a lot of time in the hearing of birds/hearing birds, and indeed all manner of animals. I feel part of a particular kind of multi-sonic/trans-somatic environment that is filled with other-than-human voicings/breathings /existences - that is always in flux, in-process, unhushed.
>
> I have always felt tremendous empathy with animals. As a child, I was appalled at the casual cruelties and unquestioning hatred and abuse of animals in the world at large. Exploitation and violation of other-than-human beings underpins our society and is embedded at every level in our h/arming hierarchies. I always felt I was no different from other animals. Having lived beside/shared life with animals, I feel this more passionately than ever now. The celebration of the transformative, merciful intelligences and energies of animals is in all my work.

On a biological level, 'kinship' is a fact: we hold much in common, particularly with other mammals. Our actions deny this as we radically shape the breeding and lives of animals, be they pets, farm animals, or wild creatures, these terms being only a few of our methods of categorisation and control that O'Sullivan references in her work, as in the gendered language of:

MARE
SOW
BITCH
(*Shaman* 15)

O'Sullivan dares to see the world differently and states this, challenging her fellow human beings to follow through the implications that we are 'no different' from animals. On one hand, then, O'Sullivan questions the dualistic divide between human and nonhuman. Yet she also, notably, uses the phrase 'other-than-human' and her poetry reflects the difference of animals as well as the differences between animals. As Cary Wolfe emphasizes, in an essay summing up the contributions of Animal Studies to ecocriticism, we are

> challenged not only by the discourses and conceptual sche-mata that have shaped our understanding of and relations to animals, but also by the *specificity* of non-human animals, their non-generic nature (which is why, as Jacques Derrida puts it, it is "asinine" to talk about *"the* Animal" in the singular [*The Animal* 31]). And that *irreducibility* of the question of the animal is linked in complex ways to the problem of their ethical stand-ing as either direct or indirect subjects of justice. (567)

Again, poetry allows the coexistence of such contradic-tions in our thinking, while probing at them and suggesting (in the final sentence) what huge gains we might make, were we to attend and 'celebrat[e] ... the transformative, merciful intel-ligences and energies of animals'. Some of O'Sullivan's poems explore a shamanistic identification with animals, endeavouring to follow them into their different spheres of inhabitation, 'Naming' entering the water as a fish, and 'Starlings' lifting up into the air:

> Lived Daily
>   or Both
>   Daily
> The Living
>   Structuring
>   Bone-Seed,
>
> Pelage,
>   Aqueous,
>
>   YONDERLY –
>   lazybed of need –
>   CLOUD-SANG
>   Tipsy Bobbles, Dowdy
>   wander (41)

Here starlings exist in a nonidealist, present-tense of subsisting, resting, flying, and singing, according to need, but the language does much more than describe. It attempts to embody another life as it may sound and feel, rather than simply to name or describe. Thus, 'Naming' ends

> this is called/
> fish. (32)

After the rush of poetry before, this feels like an abrupt sort of joke – how ridiculously inadequate naming feels, says the

dash, the full stop, the gap of anticipation before the revelation. There is the suggestion in 'Starlings' that this attempt at
empathy could affect great positive change, could 'tune me
gold'.[13] Once again, there is treasure to be found in attending
to other "critters" as Haraway would call them and experi-
menting with transformative sound and language.

Arguably O'Sullivan's embodied, sensory, corporeal 'Material Poetics', the nonvocal in her textual and performative
practice, contributes to this extension out from the merely
human, toward the 'other-than-human', challenging the long-
held conviction that it is language that distinguishes us from
other animals, suggesting indeed that there is a multiplicity
of languages in the wider world. O'Sullivan does not speak *for*
the other critters, or any individual species, but, at her most
radical, attempts to speak from within the melee of those
multiple languages, from and to a multiplicity of mammalian
selves. So it is that she inhabits the borders of the semiotic
and the semantic realms of language and both acknowledges
and questions the human/nonhuman borderline more chal-
lengingly than in statements made in interviews, for instance,
which we can quickly assimilate and hence take or leave.

Where physical bodies are present in O'Sullivan's landscapes, it is often hard to determine whether these are human
or nonhuman. Their common physicality emerges. Her work
anticipates contemporary ecocritical thinking in this: Stacy
Alaimo, for instance, argues in her groundbreaking *Bodily Natures* (2010) that it is in the bodily that we meet and realise that
we are all transcorporeally enmeshed, that all agency is in-
teractive and humans, too, are the stuff of the material emer-

gent world.[14] Throughout *A Natural History in 3 Incomplete Parts*, we find mention of body parts, but with no clear signal as to whose body is alluded to, a human, the author, a protagonist, an animal or fish or bird, even the earth itself:

> Point.of.GARDENIA/steeped.MINT.Bedum.Lash.Mantle.
> delphinium.Night.shot.
> into.the.Skin.of.
> ....
> Octave.fromz.peeled.or.lime.berry.blue.WATERING/Breast.
> SPACING.BLUE.BATHE.LANGUAGE.White
> water.Go.Water.Go.drape.Bamboo.sidling.Button.Tincture.
> Addered.Thread.Sign.Blue.teem.
> husky.Plunge.Go.Okra.PULSING.rind.pear.vein.&.Beat.Fuschia.
> Edge.tart.ice.&.BEAT. (np)

Here, in the early pages of *A Natural History*, the physicality of the text has an edgily erotic feel, with the skin and breast penetrated by natural forces and pulses, the steeping scent of flowers, the flow of water, and the beat of blood through the vein, whether it be vein of human, fish, pear, or all of these. There is a beat and a merge in the structure of single words punctuated by individual full stops, which, acting contrarily to their usual function, are designed here to draw us on through the text.

The language of *A Natural History in 3 Incomplete Parts* soon moves out of eroticism into violence, as images of slaying and mutilation multiply. The reddish brown colour of around half the text imparts bloody connotations, and the transition from

'Incomplete' to 'Most Incomplete' gains a more sinister meaning than we garner at first glance:

> slippage, mirror.
> flecked scalping (flame hue) inflict
> lemon. trace, trace bespattered clashing rib, RED Lock
> Multitude. Locks Black Bluish Black. Locks curl razor
> down womb shooting taupe rotted solvent/Tongue posse
> done w/NOW. done w/.
> done w/claw loaden keel,
> done w/FIRE down the spine,
> done w/GUT blooding down ashes.
> Ashes.
> done w/larken shot.
>
> done w/this my BODY,
> done w/this my BLOOD,
>
> done w/this my (np)

Once again, human and nonhuman signals are mixed here,
and the rib, womb, tongue, spine, and gut are not associated
with a particular mammal, though it is implied that the violence is human. This passage builds into a deathly catalogue,
each repeated 'done w/' cut off midword with a brutal slash.
The slaughtered are associated with sacrifice, as is signalled
by reference to the Christian sacrament, but it is a sacrament
shorn of its sacred qualities, a sacrament that is itself 'done w/'.
  Similarly, in 'Another Weather System' in *In the House of the
Shaman,* O'Sullivan follows a reference to 'fur' with one to 'the
disappeared' (26). This phrase, more commonly associated

with politically motivated murder on the part of military governments, here suggests individual animal deaths, and also
perhaps the disappearance of a species, as well as, of course,
the human disappeared. As such it creates a sense of "shared
subjectivity" in the face of endangered habitats, as Burnett
concludes in her essay on Clare and O'Sullivan (490). Indeed,
there are chilling moments throughout, such as '(No refund or
any/ turned or furred)' (22), in which she alerts the reader to
the irreversibility of the destruction of habitat, the possibility that species may never return, in a few bitter well-turned
words. O'Sullivan's portrayal of animal slaughter for any purpose, including consumption, is clear in the line 'beaten meats
in the head' and the references throughout to corpses and various body parts (23).

In terms of ecoethical politics, then, the phrase 'Kinship
with the Animals' poses a challenge. It implies that all are
equal, members of a family entitled to common respect, even
rights. Could it actually be *possible* to believe in shared kinship
with animals, and where would this take us? By presenting
human and animal bodies, movements, and actions in ambiguous relation to each other, O'Sullivan's writing poses these
questions. She also suggests that true 'kinship', in terms of behaviour between humans and animals, is an ideal rather than
a present reality in any context. As the extracts above imply,
she frequently exposes the 'exploitation and violation of other-than-human beings' mentioned in the interview above and
relates this to the violence in human society.

In the evocations of violence, rupture, and bleeding that
occur throughout her work, O'Sullivan frequently uses lan-

guage associated with human violence against animals, and images, too, such as the shocked face of a trapped tiger on the cover of *Unofficial Word*. More often than not, however, we cannot tell whether O'Sullivan is presenting human violence against human, human violence against animal, or animal violence against animal. Many of her pictorial images reproduce this indeterminacy. The red cover page of *States of Emergency* shows a violent image of a bleeding skull, seemingly an animal skull in the shape of a human jaw. The mixed-media piece on the front of *In the House of the Shaman* (see fig. 2) is a draped bloody cross and a panel of melted amalgamated fragments, where we glimpse a human breast and torso, a fishlike tail, and a winged figure.

The Christian emblem suggests a sacrificial kinship with animals, as the title of the piece, *An Order of Mammal*, implies. However, this title also references hierarchies of beings, the radical implication of all these representations put together being that animal suffering matters as much as human suffering. The 'h/arming hierarchies' of Western ideology and religion, which place animals below humans in terms of rights and worth, are under attack (np). *An Order of Mammal* does not place humanity at the top of the rood, but each creature in its element, bird above, human breast in the middle, and fish tail below. The multivalent punning lines 'A lied/ Be leaf' in 'Another Weather System' stress the self-deceiving nature of our convictions about the more-than-human, as well as implying perhaps that we might do less harm as a leaf or if we attempted, perhaps, to act as a leaf, subject to change (23). In a further fruitful use of multivalent language, the 'BLOOD-LINES/

on soil', which O'Sullivan follows in 'Another Weather System' and elsewhere, reflect our biological link to animals through bloodlines, yet simultaneously our role as perpetrators of violence, leaving lines of blood across the land (21).

Fig. 2. Maggie O'Sullivan, An Order of Mammal, 1987, Mixed Media, 4' × 6'

Shamanic recovery and Palace of Reptiles

In this final section, I return to the spiral of O'Sullivan's cir-
cling work to consider how her ecoethical ideas play out from
their early evocation in *A Natural History* (1985), through the
processes of *In the House of the Shaman* (1993), and finally, into
Palace of Reptiles (2003). *In the House of the Shaman* is pivotal in
that it opens with 'Another Weather System', a poem references environmental degradation and contamination of all kinds,
including 'decibel poisons' and ill treatment of animals. The
earth as a whole is under a remorseless, even endless, threat:

> All day & never – A slash, A
> scream, A smash, A burn pushed
> in
> side (each within the)
>
>   gag
>
>   BIRD OR FISH (23)

In her interview with Dell Olsen, O'Sullivan writes that 'the
place of the page' is 'A place of damage, savagery, pain, silence:
also a place of salvage, retrieval and recovery'. In 'Kinship with
Animals', the second part of *In the House of the Shaman* (a book
named after a Joseph Beuys drawing), we have already seen
a more positive, at times shamanistic, engagement with the
more-than-human world, a suggestion that change is possible. This section has 'Of Mutability' at its heart, and indeed the
heart of the whole book, and it culminates in the exuberant

wing-embraced 'Wingsunsong for Cobbing'. O'Sullivan cites Beuys in her epigraph to 'Kinship with Animals':

> To stress the idea of transformation and of substance.
> This is precisely what the Shaman does in order to bring
> about change and development: his nature is
> therapeutic. (28)

Whereas moments of 'dance' and 'weave', akin to the movements the Dancer will make in *Palace of Reptiles,* are rare in 'Another Weather System', 'Kinship with Animals' celebrates the familiarity and mystery of the nonhuman in the 'DANSING' of 'Bog Asphodel Song' and the sonic joyful energy of poems such as 'Starlings' and 'Naming' (24, 33). In the final part of *In the House of the Shaman,* 'Prism and Hearers', O'Sullivan attends to the 'fugitive gods', 'Hill Figures', and bodily transformation (49-68), culminating in a collage embedding text in insect and feather images – in other words, materiality suggesting Irish shamans' cloaks (Sheppard, 173).

In *Palace of Reptiles,* O'Sullivan deepens her evocation of ritual as therapeutic and material engagement with the more-than-human world, but also critically examines how we might use this. Human mythologizing of landscape can be viewed as anthropocentric appropriation, but, although she references Irish mythology in particular, O'Sullivan does not mythologise through closed narratives, ancient or new. Rather, she uses the idea of ritual to explore engagement with the more-than-human world. Her consciousness of contemporary environmental degradation confirms that this work cannot be accused of escaping to a 'timeless', 'natural' or 'primitive' age. Two poems, presented as a pair, are particularly illustrative of O'Sullivan's complex, ambivalent, and challenging use of ritual. Their titles, 'Narcotic Properties' and 'Theoretical Economies', are redolent of the realms of science, medicine, and economics but are basically facetious, both poems dealing in an understanding that goes beyond what is explicable or culturally acceptable.

Both poems extend beyond the written text, making reference to performative actions and visual images, some of which are included in the book-space as a whole. In these poems, O'Sullivan seems to be creating rituals to reenact and perhaps recuperate human environmental crimes, especially in the treatment of animals.

Peter Manson, in his review of *Palace of Reptiles,* writes, 'It's never really possible to decide whether the actions are therapeutic rites, performances in a gallery space, or the private rituals of an obsessive-compulsive' (32). They may be all that Manson mentions, but they are also more. His emphasis here is on the author (her therapy, performance or ritual), but if we read with an eye to the wider world, the poems become accusations of complicity and instructions for acknowledgment, if not partial redemption. It is of key importance that both poems directly instruct the reader to perform ritual actions. The poetic persona is more than protagonist here: she becomes a shaman or leader of the community.

'Narcotic Properties' opens with the following lines:

PLACE A SMALL PALE-CREAM BOWL (TO SIGNIFY
abundance)
on the table-top in front of you.

Wash THE FOLLOWING LEAD ANIMALS: trout, dog,
tiger, owl, moth, wolf, cat, drake, bat, fish,
pig, dolphin, buffalo, bee, scorpion, snake,
eagle, crab, goat, salmon, turtle. IT IS IMPORTANT
TO WASH THESE LEAD ANIMALS WITH songerings-a-rung,
a-chant, a-roughy
unsway
& stirs
(Palace 16)

The poetic persona goes on to instruct the reader/listen-
er in a ritual in which the sequence of ritual actions is clear,
though its meaning is not. The animals are washed, dried, and
placed on a cloth, which then begins to bleed and is lit with a
match. A scene follows in which coloured flags and lights seem
first to be erected and then to fall to the floor. The scrappiness
and commonality of the objects O'Sullivan references here,
and actually uses in her performative work, too, are in keeping with her self-described 'ASSEMBLAGES, after Kurt Schwitters who made superb use of the UN – the NON and the LESS –
THE UNREGARDED, the found, the cast-offs, the dismembered
materials of culture' (67). Though this assumes that the lead
animals are old metal toys, while there are, of course, many
other meanings of lead in the form of both verb and adjective.

Is the poem a celebration or a lament? It is both, moving
between tenderness (caressing, talking, singing) and containment, even violence ('PENS RINGWAYS, CAGES') in its enact-

ment of human behaviour toward animals (18). At one level, it is a ritual, seemingly serious in its instruction to the reader to undertake these actions. At another, it is a mock or ironic ritual scattered with lines such as 'Dry each lead animal in turn with a portable hairdryer' and 'THERE MUST BE NO OVER-CROWD-ING' (17). Manson is right in that we simply do not know how to read these poems and, therefore, they are 'deeply, humorous-ly, disquieting constructions' (32). This is not a ritual in which the human being escapes the realities of the everyday into a depoliticised celebration of the glories of humans and ani-mals existing in kinship. Rather, as the title might suggest, the hope is that through these actions, in which representations of animals are engaged with, the participant is awoken from his or her 'narcotic' numbness or insensibility into greater knowledge and awareness.

In ecoethical terms, the bleeding of the cloth and the ref-erences to the keeping of animals in overcrowded enclosures and to their transportation all call attention to violence and mistreatment. One reading of the word 'Properties' in the title is to see it as invoking the traditional Judeo-Christian assump-tion that humans own animals as property. Yet the piece is still haunted by a more positive picture of kinship, past or future, and ends with a reference to a blue light of 'KINSHIP', though it be 'OOZED OUT OF SHAPE' (18). Similarly, the last word of 'The-oretical Economies' is 'weeps': the 'theroretical economies' of our culture have translated not into a thriving kinship among the living and growing, but into a scene of massacre and dev-astation. In both poems, however, there is also a sense of rit-ualistic artistic practices attempting to claw back or create anew different ways of living and being, ways that might be ecoethical.

In challenging readers directly with a list of instructions to act, O'Sullivan dares them to acknowledge and respond to the state of the world they live in. In performance this challenge is even more direct. In interview, Dell Olsen and O'Sullivan discuss her performance techniques, involving for instance the poet walking among the audience, weaving red net among them, or laying pages of text on the floor and using recordings and slides, as well as her own voice. O'Sullivan uses the word 'IMPLICATION' to characterise this form of poetry reading. She refers in part to 'page as place', the embodiment of text in space. 'Implicate', however, has further meanings: 'To intertwine; To entwine, entangle;... To involve; to bring into connection with' (OED), and this is precisely what O'Sullivan aims to do in her performances – to draw the spectator into an intimate relationship with the material they are hearing.

O'Sullivan's epigraphs to 'Doubtless', the poem we began with here, are from Tom Lowenstein's *Ancient Land, Sacred Whale: The Inuit Hunt and Its Rituals*. Lowenstein also uses the word 'implication', in this case referring to the 'knots, whorls, vortices of human implication in the landscape' (Palace 31). Here 'IMPLICATION' refers to our place in the wider world we inhabit, as well as our place as the audience for a performance of poetry. One should surely lead to the other. O'Sullivan's work can return us to the world more aware of our own and our society's tracks in the land and the fact that these affect more than ourselves. As she says in her interview with Olsen, 'As we hear, we also are heard in an intertwining of potential

exchange of hearing-(being)-heard of other-than-(as well as
human)-sentience'. In struggling to reach out into a radically
imaginative, even shamanistic, voicing of and identification
with the more-than-human world, she strives to unite tongues
with moor, bird and mammal, to move away from anthropo-
morphic perspectives and erode simplistic binary divisions.
Above all, it is her extraordinary language, her written forms,
visual pieces, and performances that she stimulates her readers to new and radical imaginings within and about environment.

End Notes

[1] For a discussion of Bunting as predecessor of British experimental landscape poetry, see Tarlo, "Radical Landscapes."

[2] See Middleton for a wonderful and witty discussion of neologism in O'Sullivan.

[3] See Scott Thurston's interview with Maggie O'Sullivan (Thurston, 2004) for discussion of the method of compilation of this text.

[4] O'Sullivan dedicates the first and last parts of States of Emergency to Basil Bunting and Joseph Beuys, respectively. Both are cited as sources for her piece 'riverrunning (realisations', printed at the end of Palace of Reptiles.

[5] See the Eclipse archive at http://eclipsearchive.org for reproductions of some of these early texts.

[6] In the spirit of Beuys, O'Sullivan also asserts that life and poetry should not be seen as separate, as another answer to Thurston demonstrates: 'Coming back to language, I planted eight hundred and fifty trees, which to me is an act of poetics, just as my language is an act of poetics. And I would say the two have equal presence for me as a practice of poetics: of making meaning' (12).

[7] See the poems of Waterfalls, some of which appear in O'Sullivan et al.

[8] I am working within the 'ecopoetic' field here but am resisting rehashing the debates about terminologies that I have discussed elsewhere, particularly in the introduction to my anthology, *The Ground Aslant.*

[9] Observant readers will note I have used 'more-than-human,' 'nonhuman' and 'other-than-human' in this essay. The former is my preferred term and especially for the wider plenitude of O'Sullivan's work; 'nonhuman' is used to denote binary thinking about the human relation to other creatures; 'other-than-human' is a

term O'Sullivan herself uses in interview and is only used to refer to that usage.

[10] Reprinted in O'Sullivan, *Body of Work*, 61.

[11] See Sheppard's *When Bad Times Made for Good Poetry* for a discussion of the origins and nature of this text in the context of the 'poetics of innovation'.

[12] An earlier, shorter version of *A Natural History in 3 Incomplete Parts*, titled *An Incomplete Natural History*, was published by Writers Forum in 1984 and reprinted in 1994. A facsimile of this edition is available at:

http://eclipsearchive.org/projects/NATURAL/html/pictures/nh0007.html. Peter Manson gives more detail and analysis about its material history and collage nature in his essay on the work.

[13] Tuma gives an example of how critics interested in linguistically innovative poetry can fail to see that such poetry is about any more than language: "For all the animals and animal parts littering her work, the starlings enter this poem only as a lure, occasion for a bravura display of language, the referent for a hyperexpressionistic linguistic excess" (23).

[14] Mandy Bloomfield has also made a more detailed series of connections between Alaimo and O'Sullivan in "Unsettling Sustainability".

Works Cited

Alaimo, Stacy. *Bodily Natures: Science, Environment, and the Material Self.* Indiana UP, 2010.

Andrews, Bruce, and Maggie O'Sullivan. *EXCLA.* Writers Forum, 1993. Facsimile available at http://eclipsearchive.org.

Bate, Jonathan. *The Song of the Earth*. Picador, 2000.

Bennett, Jane. *Vibrant Matter: A Political Ecology of Things*. Duke UP, 2010.

Bloomfield, Mandy. "Maggie O'Sullivan's Material Poetics of Salvaging in red shifts and murmur." *The Salt Companion to Maggie O'Sullivan*, ed. Chris Hamilton-Emery, Salt Publishing, 2011, pp. 10-35.

_____. "Unsettling Sustainability: The Poetics of Discomfort." *Green Letters*, vol. 19, no. 1, 2015, pp. 21-35.

Bryson, J. Scott., *Ecopoetry: A Critical Introduction*. University of Utah Press, 2002.

Buell, Lawrence. *The Environmental Imagination*. Harvard UP, 1995.

Bunting, Basil. *Collected Poems*. Oxford UP, 1977.

Burnett, Elizabeth-Jane. "Sounding the non-human in John Clare and Maggie O'Sullivan." *John Clare Society Journal*, Jun 2016, Issue 35, pp.31-52

Campbell, SueEllen. "Magpie." *Writing the Environment: Ecocriticism and Literature*, edited by Richard Kerridge and Neil Sammells, Zed, 1998, pp. 13-26.

Durand, Marcella. "The Ecology of Poetry." *Ecopoetics*, no. 2, 2002, pp. 58-62.

Garrard, Greg. "Heidegger Nazism Ecocriticism." *ISLE*, vol. 17, no. 2, 2010, pp. 251-71.

Haraway, Donna J. *Primate Visions*. Psychology, 1989.

_____. *When Species Meet*. U of Minnesota P, 2013.

_____. *Staying with the Trouble: Making Kin in the Cthulucene*. Duke University Press, 2017.

Hart, George. "Postmodernist Nature/Poetry: The Example of Larry Eigner." *Reading under the Sign of Nature: New Essays in Ecocriticism*, edited by John Tallmadge and Henry Harrington, U of Utah P, 2000, pp. 315-32.

Head, Dominic. "The (Im)possibility of Ecocriticism." *Writing the Environment: Ecocriticism and Literature,* edited by Richard Kerridge and Neil Sammells, Zed, 1998, pp. 27-39.

"implicate." *OED Online,* Oxford University Press, March 2020, www.oed.com/view/Entry/92474. Accessed 1 April 2020.

Lowenstein, Tom. *Ancient Land, Sacred Whale: The Inuit Hunt and Its Rituals.* Harvill, 2003.

Manson, Peter. "A Natural History in 3 Incomplete Parts." Hamilton-Emery, pp. 71-79.

————. *Review of Palace of Reptiles,* by Maggie O'Sullivan. The Paper, no. 7, Nov. 2003, pp. 29-37. http://poetrymagazines.org.uk/magazine/print770d.html?id=28540.

Marsh, Nicky. "Agonal States: Maggie O'Sullivan and a Feminist Politics of Visual Poetics." *Hamilton-Emery,* pp. 80-96.

Middleton, Peter. "'Ear Loads': Neologisms and Sound Poetry in Maggie O'Sullivan's Palace of Reptiles." *Hamilton-Emery,* pp. 97-122.

O'Sullivan, *Maggie. Body of Work.* Reality Street, 2006.

————. Interview with Charles Bernstein. "Writing Is a Body-Intensive Activity: Close Listening with Maggie O'Sullivan." *Jacket* 2. 4 Dec. 2013. http://jacket2.org/interviews/writing-body-intensive-activity.

————. Interview with Dell Olsen. "Writing/Conversation: An Interview by Mail, November-December 2003." *How2 work/book.* 2 Feb. 2004. www.asu.edu/pipercwcenter/how2journal/archive/online_archive/v2_2_2004/current/workbook/writing.htm.

————. *In the House of the Shaman.* Reality Street, 1993.

————. *A Natural History in 3 Incomplete Parts.* 1985. Magenta, 1997. Reprinted in black and white in *Body of Work.* Facsimile available at http://eclipsearchive.org.

————. *Palace of Reptiles.* Gig, 2003.

_____. *States of Emergency.* International Concrete Poetry Archive Publication No. 11, 1987. Facsimile available at http://eclipsearchive.org.

_____. *Un-assuming Personas.* Writers Forum, 1985. Facsimile available at http://eclipsearchive.org.

_____. *Unofficial Word. Galloping Dog,* 1988. Facsimile available at http://eclipsearchive.org.

_____. *Waterfalls.* Reality Street, 2012.

O'Sullivan, Maggie, David Gascoyne, and Barry MacSweeney. *Etruscan Reader III.* Etruscan, 1997.

Perloff, Marjorie. "'The Saturated Language of Red': Maggie O'Sullivan and the Artist's Book." Hamilton-Emery, pp. 123-35.

Rigby, Kate. "Earth, World, Text: On the (Im)possibility of Ecopoiesis." *New Literary History,* vol. 35, no. 3, 2004, pp. 427-42.

Sheppard, Robert. *Far Language: Poetics and Linguistically Innovative Poetry, 1978-1997.* Stride, 1999.

_____. "The Performing and the Performed: Performance Writing and Performative Reading" *How2,* vol. I, no. 6, 2001. www.asu.edu/pipercwcenter/how2journal/archive/online_archive/v1_6_2001/current/index.html.

_____. *When Bad Times Made for Good Poetry: Episodes in the History of the Poetics of Innovation.* Shearsman, 2011.

Chun, Sehjae. "'An Undiscovered Song": John Clare's "Birds Poems'." *Interdisciplinary Literary Studies: A Journal of Criticism and Theory,* 6:2, 2005

Soper, Kate. *What Is Nature? Culture, Politics and the Non-human.* Blackwell, 1995.

Tarlo, Harriet. *The Ground Aslant: An Anthology of Radical Landscape Poetry.* Shearsman, 2011

————. "Radical Landscapes: Contemporary Poetry in the Bunting Tradition." *The Star You Steer By: Basil Bunting and British Modernism*, edited by James McGonigal and Richard Price, Rodopi, 2000, pp. 149-83.

Thurston, Scott. "Emerging States." *Poetry Salzburg Review*, no. 6, 2004, pp. 6-14.

Tuma, Keith. *Fishing by Obstinate Isles: Modern and Postmodern British Poetry and American Readers.* Northwestern UP, 1998.

Wolfe, Cary. "'Human, All Too Human: "Animal Studies" and the Humanities." (2009). *Publications of the Modern Language Association of America.*, 124(2): 564-575.

Fractured Environments in Susan Howe's *That This*

SARAH NOLAN

In 2011, Susan Howe released the audio project "Frolic Architecture," a collaboration with composer David Grubbs in
which words, syllables, ambient noise, music, and poetry are
audibly layered into a performance that both fractures and
enmeshes the historical, personal, textual, and material environment of the poetry. The work emerges from Howe's mixed-
genre book *That This* (2010), the second section of which provides the poetic core of the collaborative performance as it
highlights the highly fragmentary and intensely visual nature
of the text through appropriated material, which is arranged
on the page so that lines of poetry overlap and words are cut
off in visual tears, making the poems appear as scraps of paper
torn from a book. The poetry is framed by six photograms by
James Welling, and much of the text emerges from the writing
of Hannah Edwards Wetmore, which the poet encountered
on a visit to the Beinecke Library. Yet, as Howe comments in

a 2011 performance at Harvard University, through her work with Grubbs, she has "moved away from a standard, set narrative into something much more fractured and abstract," and the collaboration made her realize that her "words could be further broken apart into a kind of sound that is itself a sort of music but that still does have some narrative" (Grubbs and Howe). Howe's identification of a distinct narrative in the performance is a stretch for most listeners, who struggle just to pull out the words from the layered recordings of broken vowels and consonants mixed with Grubbs's other sound elements. The performance, though, serves an important function in understanding Howe's complex and sometimes even incomprehensible poetics by drawing listeners into a spatial experience in which divergent sound patterns and poetic lines fill the room simultaneously and emphasize the disjunctive and multiplicitous reality of lived experiences.

Despite the predominance of space as a shaping force in *That This* and "Frolic Architecture" and the spatial undertones of much of Howe's other writing, ranging from *Articulation of Sound Forms in Time* (1987) to *The Midnight* (2003), her work teeters perilously on the periphery of ecocritical studies – a field historically dismissive of poetry and continuously at odds with the sentiments of contemporary poetry. To better situate Howe's work in relation to ecocriticism, I will consider how triangulating her work with ecopoetic discourse and recent developments in ecocritical theory – namely, material ecocriticism – reveals the fractured environments that her poems create and the agential power of words in shaping those spaces. Working through the text of That This in relation to

Howe's collaborative performance of the "Frolic Architecture" section of the book, I will show how the poet's work reframes ecocritical understandings of space through her disjointed textual environments and her emphasis on text's ability to shape experiences in the world.

Ecocritical Readings of Howe and New Theoretical Approaches

Although the visual elements in Howe's work imply a deep interest in space, her poetry has received little attention from ecocritics. The two prominent exceptions are Scott Knickerbocker's discussion of Howe in the concluding chapter of his 2012 book, *Ecopoetics. The Language of Nature, the Nature of Language,* and Mandy Bloomfield's 2014 article "Palimtextual Tracts: Susan Howe's Rearticulation of Place." Bloomfield's article, which provides a concise overview of the extensive body of work on Howe's connection to landscape and its tenuous engagement with ecocriticism, oscillates around an understanding of Howe's poetry as contradicted in its engagement with a tradition of place-based writing and an understanding of the inability to ever achieve an unmediated interaction with the physical world (667-68).

Bloomfield rightly recognizes the materiality of Howe's work, but only tentatively in her concluding paragraphs. I will attempt to pick up her concluding charge to consider what role new materialist ideas might play in reading Howe's poems by specifically locating material agency in the text.

In Knickerbocker's limited treatment of Howe, he contends that the poet presents "language as itself wild, part of nature broadly conceived. Like (the rest of) nature, language comes before us and exists beyond us; language shapes the writer—indeed, calls the writer into being—as much as the writer shapes language" (175). Although Knickerbocker's understanding of language's power is compelling, I see Howe's text as significantly more effectual than simply shaping the writer. Expanding on Knickerbocker, I view Howe's work as making language active in shaping how individuals experience history, culture, and space. In other words, Howe's text shapes not only the writer but the reader as well.

My departure from Knickerbocker, though, is perhaps best illustrated by considering how his central claim is put into action in relation to Howe's work. Knickerbocker rightly emphasizes Howe's interest in sound, which reflects the concept of "sensual poesis" that he posits as it attempts to re-create the experience of the senses in real-world encounters. His discussion revolves around Howe's 1987 *Articulation of Sound Forms in Time,* one of the poet's few books inspired by an experience in the wilderness.[1] In selecting a text that revolves around natural experiences, Knickerbocker falls back on traditional notions of nature. While his argument interestingly points out the importance of sound for Howe, he also posits that the poem employs experimental poetics to mimic the wilderness experience of the book's central historical figure (173). My interest in Howe is unhinged from conventional con-

ceptions of nature because Howe's work is not primarily engaged with those elements of the world. Although some wild
images and encounters appear sporadically in her texts, they
are couched within self-reflection, formal experimentation,
and language play—the aspects of her work that, to me, are
particularly ecopoetic because they capture the radical in-
authenticity of all experiences by emphasizing the holes in
human perception and linguistic expression. By exploring the
boundaries of language and form within a self-aware textual
space, Howe's work achieves an environment of its own, one
fractured but true to the multiplicitous state of life in the contemporary world.

Reading Howe as an ecopoet requires a reconsideration
of what constitutes an environment. For ecopoetics, as for
ecocriticism, the movement away from traditional ideas of
environment and toward the new materialisms that integrate nature and culture demands the examination of new
types of texts that embrace what Donna Haraway terms "na-
turecultures," a term that indicates "the implosions of the discursive realms of nature and culture" (105). Haraway's "natu-
recultures" make it impossible to view agency as something
uniquely human. Rather, as the boundaries between culture
and nature break down, and the two worlds intertwine and
equalize, everything in an environment must be granted
agency. Serenella Iovino and Serpil Oppermann's 2012 article,
"Theorizing Material Ecocriticism: A Diptych," provides a help-
ful overview of material ecocriticism in which Oppermann
posits that in material ecocriticism, "the natural and the cultural can no longer be thought [of] as dichotomous catego-

ries. Rather, we need to theorize them together, and analyze their complex relationships in terms of their indivisibility and thus their mutual effect on one another" (462-63). Through material ecocriticism, then, scholars rethink subject-object relationships in order to acknowledge and engage with the "other bodies" in the world on a meaningful level. As agency dramatically expands, conceptions of environment are radically altered.

Instead of looking for traditional environments or connections to nature, new ecopoetic approaches examine texts that acknowledge and even foreground that "matter and discourse shape and interfere with each other" (Iovino and Oppermann 454). In other words, texts are part of the physical spaces they project and, even further, work to shape those spaces. Creating naturalcultural spaces on the page demands that writers reexamine their use of language and form while both foregrounding and embracing their medium's limitations. As scholars such as Oppermann and Stacy Alaimo theorize the breakdown of distinct boundaries between the human body and surrounding spaces, what is considered "environment" rapidly changes, and, I argue, what constitutes a valid site for ecopoetics similarly changes. In reading Howe's work through the lens of ecopoetics, such a shift is essential because her work is not centered around traditional environments or natural elements. While *That This* is seemingly devoid of a clearly identifiable place, Howe constructs a textual environment that, while fractured, is as tangible and valuable as the physical world.

Howe's Fractured Environments

The fractured spaces of Howe's poetry are most readily apparent through the dramatic formal structures she employs, which are often marked by fragmented words, careful use of page space, and multimedia elements from her own life, historical documents, or other artists. While she is often considered a language poet, Howe is unique in the contemporary poetry world because her background, including her education and early work in visual arts, inspired her recent shift toward a somewhat radical multimedia format. Formally trained as a painter, Howe relished the possibilities presented by collage but also acknowledges the importance of sensory experience. As she explains in an interview with Lynn Keller, she views collage as "a way of mixing disciplines" in her own work, and she recalls that even during her time working in the visual arts, she viewed the verbal titles of other artists' work as formative in her interpretation of the piece's visual elements (qtd. in Keller 4).

Howe's recognition of the power of language within visual art led her to enact precisely this kind of blending within her own work, first creating art books where she began developing her formal mixed-media methodology and eventually building material environments out of text.[2] She explains that eventually, her art consisted of "making environments—rooms that you could walk into and be surrounded by walls, and on those walls would be collage, using found photographs (again a kind of quotation)." She continues, "Then I started using words on the walls and I had surrounded myself with words that were really composed lines" (qtd. in Keller 6). Howe's poetry, then, originated in space—collaged environments, made of poetry, photographs, sounds of words, and individual experience.

Her self-identified roots in installation art help clarify her relationship to ecopoetics and, further, her decision to collaborate with composer David Grubbs to create a spatial experience for her work, which clearly emphasizes layering and fracturing. While such multiplicitous environments are foregrounded in the audio version of "Frolic Architecture," it is important to consider its original textual iteration in *That This*. In the book, the poet creates an environment much like those of her installation past by infusing multiple media into the reader's experience. The first section of the book, "The Disappearance Approach," is written entirely in prose as it jumps between her recollection of her husband's death, memories of his life, quoted material from Sarah Edwards and her sister-in-law Hannah Edwards Wetmore, and the poet's responses to that material as she explores the Jonathan Edwards papers at the Beinecke. Section two, after which the audio collaboration with Grubbs is named, enacts a radical shift in formal structure. The poems in "Frolic Architecture" are mostly collage poems, in which passages appear to be torn directly from the original texts that the poet encounters in the archive. The highly visual poems are distinguished by their layered, multidirectional texts, varied fonts, and use of white space. Such graphic elements are enhanced by the section's integration of six full-page photograms by James Welling. The third and title section, "That This," which includes only thirty lines of poetry, shifts back to a more traditional formal structure and focuses

subtly on self-reflexive commentary on language and experience.

The connection between *That This* and Howe's background
in visual and installation art is apparent in her conceptualiza-
tion of the book as a sort of experience in itself. In its three
distinctive parts, the book presents multiple perspectives on
one subjective experience—the poet's reading of the papers at
the Beinecke. The moments with the papers are infused with
memories, reflections, feelings, sensations, extracted passages, and poetic responses, which are reflected in each of the
sections quite differently. In the first section, the poet expresses the multiplicity of her time in the library through radical
leaps from one topic to the next, but grounded in the papers
she encounters in the collection. For instance, she recalls the
moment of her husband's death, concluding the paragraph,
"Starting from nothing with nothing when everything else has
been said" (11). The line reflects the poet's grieving response to
the loss of Peter, her husband, as a bleak new path. The sentiment is picked up in the following paragraph, which begins
with a quotation from Sarah Edwards: "O My Very Dear Child.
What shall I say? A holy and good God has covered us with a
dark cloud" (11). The connections between the quoted material
and personal reflection reveal the archive's material presence
for the poet. The poet's encounter with the papers in the library is accompanied by her reflection, cognitive wandering,
and memories of her late husband. The experience of the library, then, is not a purely material one. In Howe's prosaic iteration of the moment, the juxtaposition of material and nonmaterial elements, through quoted words from the papers

and the poet's own subjective response, demonstrates the archive's agential power. Although the passages in the section appear unrelated, each paragraph captures another aspect of the poet's experience, all of which is inspired and shaped by the material text in the Edwards collection. In other words, the moments in the text become powerful or agential because they attain the capacity to shape what follows. The poet responds in various ways based on the text she encounters and the thoughts or feelings those readings inspire. Her recollection of her husband's death is brought on by the words of Sarah Edwards, a connection that inspires contemplation and reflection of the poet throughout the section.

The influence of nonmaterial aspects of lived encounters and the importance of environmental experience are most apparent in the second section of *That This* and reflected in the corresponding audio project. In the book, layering is literal on the page, which creates a clear visual of the agential power of the texts in the Beinecke. The texts that the poet encounters physically dominate and shape the form and content of the book's second section, indicating the influence of the Edwards archive over the poet's expression of her own life. As the first section indicates, the poet's memories, thoughts, feelings, and comments are directly inspired by the material she encounters in the library. In this way, the material papers in the collection hold some power over the poetry, inspiring and shaping the poet's physical and cognitive responses. Alongside the archive's power to shape the poem, Howe's use of the material in the second section of the book gives the text another element of agency. Although the poet imposes herself on the text through selection, placement, and order of the snippets, the original text maintains a sense of power in shaping how readers experience the finished poems and, perhaps more important, how readers of Howe's poems interpret, understand, or interact with the Edwards archive itself. Ultimately, materiality emerges in two ways in *That This* – first, through the material influence of the archives themselves on the ways in which the poet recounts her lived experiences on the page, and second, through the real-world effect of readers encountering the Edwards papers within the fractured space of Howe's collaged poems.

The power of the Edwards papers remains prominent and constantly underlies Howe's original contributions. In "Frolic Architecture," the lines of text are placed on top of one another, at various angles, and often overlapping. The text looks like an image of a torn paper scrap scanned onto the page. By collaging snippets from the archive, the poet is visually fracturing her experience in the library. In a sense, the fragments of text are reminiscent of the fragments of thought that she introduces in the first section of the book. While "The Disappearance Approach" articulates how lived experience fractures into material and nonmaterial elements – ranging from memory to cognitive wandering – through radical shifts between prose passages, "Frolic Architecture" visually depicts the fractured and multifaceted ways in which human beings always interact with the material world. Such fragmentation is not uncommon in Howe's poetry, and it often occurs in response to historical figures with whom her work frequently interacts. Paul Naylor, who specifically focuses on Howe's engagement with historical figures, points out that "the events in history are not, despite appearances, woven into a seamless, hypotactic narrative; information has been, intentionally and unintentionally, left out. Without that information, without the normal hypotactic connections between elements within the narrative, we become, as Howe puts it in another poem, "'Lost in language,' although 'we are language'" (61). The poet's collage of the original text and corresponding disjuncture of her own experiences that are reflected in the book may indicate her proposal that experience, like history, is inexact.

Within the fragments that she draws from the papers in the library, Howe presents a formal statement on the fractured nature of experience; however, she also guides the reader through her poetic methodology within the content of the poems.

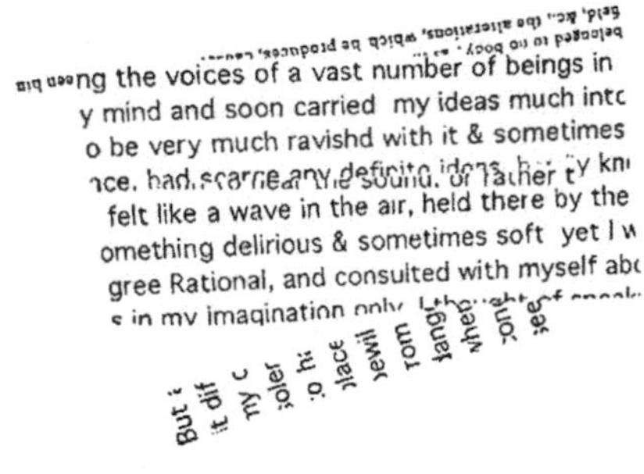

Fig. 1. "the voices of a vast number" from Susan Howe, That This, New Directions, 2010, 57.

Visually, this section of the poem appears more like a piece of visual art than poetry. The borders of the text are irregular and torn, and they often include fragments of words. It gives the reader the sense that the text was torn from the original material in the Edwards archive, but it also employs the sentiments of appropriation artists and writers, ranging from the renowned Marcel Duchamp to the more recent conceptual poetry of Kenneth Goldsmith, in which material is directly adapted from an original source and recontextualized in a new space. In the work of these artists, as in Howe's poetry, the content is adapted from another source, but it is entirely original in its new context. In fact, the found poetry of the book's second section carries forward the sentiments established in the entirely original prose of section one, implying the poet's power to shape appropriated material. Aside from the radical formal structure of the "Frolic Architecture" section, the poems comment on the layered or multiple aspects of a lived encounter—in this case, the poet's experience of the Beinecke. Through both its form and content, the poem implies the limited capacities of words, the gaps inherent in human experience, and the subjective sensations and feelings that accompany a real-world encounter.

Reading past its fractured structure, one finds that the content of the poem similarly depicts an environment of multiplicity. The poem recounts "the voices of a vast number of beings in [the speaker's] mind" who "carried [her] ideas." The fourth line of the most prominent section embodies its own sentiment, as it appears to read, "had scarce any definite ideas." The words, though, are cut off, leaving the reader uncertain and the poem without any "definite ideas." All elements of the text are so deeply fragmented that it provides an unsettling reading experience but also, more important, a sense that the fractured nature of the text belongs. It creates an experience for the reader, a moment that captures the multiplicity of the poet's own encounter with the archives. As section one demonstrates through prose, the poet's mind leaps from topic to topic, some connected to the Edwards material she encounters and some cognitive tangents or memories. Such is the reality of human experience – a fragmented encounter with the world that somehow attains a sense of cohesion within the mind. Howe's poem embodies the reality of multiplicity, confusion, and chaos. In a sense, her poem does precisely what it claims to do as it allows the reader to experience it "like a wave in the air, held there by the / something delirious & sometimes soft." Leaving uncertainty in the incomplete "something," these lines encapsulate the environmentality of Howe's poetics.

Through her fracturing of the encounter, she creates an environment where text shapes experience—for both the poet and the reader. Her words are shaped by the material she reads, an influence that grants the original material text agential power. Secondarily, though, Howe's reconstituted text attains a sense of agency in itself. The puzzling gaps that the incomplete words create in the reading process shape the experience of the text and, in a sense, make it a lived space in itself as the reader struggles to attach meaning to seemingly meaningless fragments of language.

The fractured textual space of "Frolic Architecture" is emphasized in the poet's mixed-media approach and continually reflected in the language of the poems. The photograms that accompany the text, the work of James Welling, lend cohesion to the fragmented textual elements of the book. The photograms depict dreamlike landscapes in black-and-white imagery. They appear almost like slides under a microscope, blending shadows and light in various patterns. Each image fills a page of the book. The images literally embody the language of the poems.

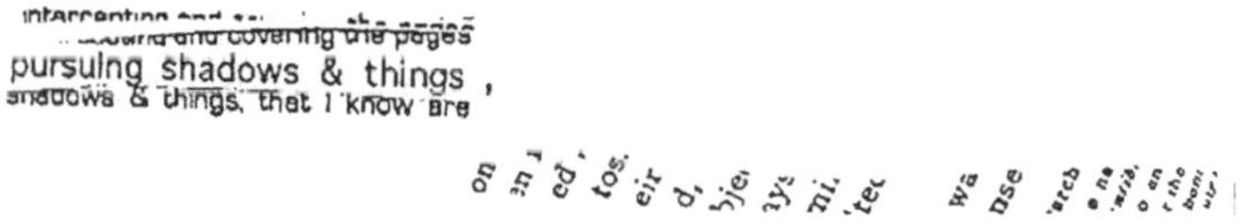

Fig. 2. "pursuing shadows & things" from Susan Howe, That This, New Directions, 2010, 62.

The words of the poem are obscured by a hand-drawn line, which appears to leave the poem reading, "intercepting and / and covering the pages." The fragments of letters and illegible words are not to be forgotten, for they are the core of the lines. The "intercepting" that the poem points toward both applies to the photograms, which intersect the poems and cover the pages, and extends into the poet's process, as her marks intercept and break the first two lines of text. Howe's inclusion of a drawn line that runs through the text creates uncertainty and disconnection. As throughout much of the section, the reader must piece together meaning from the fragments of words that remain. In a sense, the reader ends up "pursuing shadows & things" in his or her search for meaning in the remnants of text. The repeated line recurs with an alteration, "shadows & things that I know are." The shadows again take up two meanings in the book – first, there are the shadows of words that remain behind, unclear, and only a reflection of an original cohesive text; second, the images are shadows of the poems, as they reflect in visual form the sensations of the textual space.

The concept of the photograms in the second section of *That This* is, arguably, the inspiration for the poet's collaboration with Grubbs on the audio CD project "Frolic Architecture." The album embodies the sentiment of the text by embedding the listener within an experience that is made up of layers of sound, music, fragmented words, and complete poetic lines. In a sense, the audio aspect of the project acts similarly to the photograms in the text but takes on a much more dominant role. The audio elements draw the listener into a multifaceted experience of the poetry, one that is not primarily words but consists of other sensory aspects of the text. Grubbs's sounds on the project capture the sensations of Welling's photograms through a dreamlike simplicity of sounds ranging from a long, extended drone of background noise to the growing and fading sound of a rattle to the crunch of dead foliage. His soundtrack is constantly changing, embedding instrumental sounds within recordings of Howe's voice. Throughout the project, audio tracks of Howe reading complete sections of the poem are layered with tracks of the incomprehensible fragments on the page. Interestingly, those parts of the text that appear to be unreadable are included in the project and captured as a layering of sound. The fragments of words create a cacophonous musical track, but they add a sense of multiplicity or simultaneous thought to the poems. In other words, the audio project succeeds in emphasizing layering within experiences even more so than the radical formal structure of the poems on their own. Although the book clearly includes fragments from the Edwards archive alongside more complete sections of text, the impulse of the average reader is to look past the illegible fragments and locate complete, readable poems in the larger sections. The audio project, though, makes such selective reading impossible. Instead, the layers remain distinctively intact, forcing the reader to acknowledge and embrace the multifaceted experience.

The concept of the audio collaboration between Howe and Grubbs is profoundly environmental. In fact, the project creates an environment much like Howe's origins in installation art as it defines a multifaceted experience in space. Chloë Bass, a conceptual artist and columnist for the art and culture website Hyperallergic, describes the spatial aspects of the collaboration. After witnessing Howe and Grubbs perform "Frolic Architecture" live, Bass identifies within it the creation of "a space where both players [Howe and Grubbs] are 'co-present,' where each layer remains visible and active on its own. That simplicity requires that each element have its own clarity and value, both tonally, through a careful non-mixing of frequencies, and linguistically, through limited words. Layering is a way to hold moments together, a captivating rather than obfuscating force" (par. 4). Through the layered audio tracks, including Howe's poems and their origins in the Edwards archive, the audio project captures the fragmentation and uncertainty of lived experience. Such a response to the collaboration is reflected by the poems in the book.

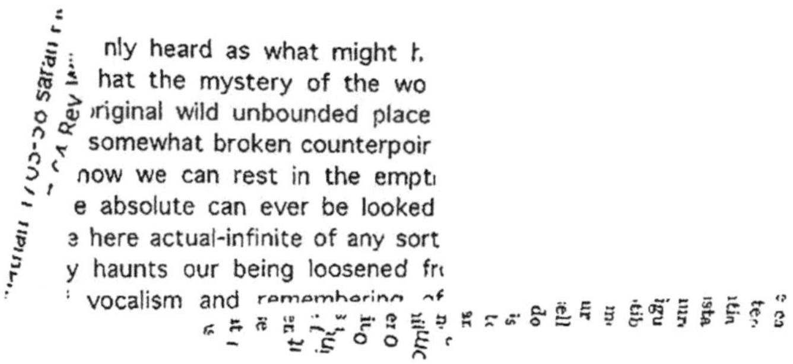

Fig. 3. "heard as what might" from Susan Howe,
That This, New Directions, 2010, 75.

The large, more readable fragment of this poem echoes the sensations of its recorded counterpart. The audio recording captures the chaotic, unpredictable, and multiplicitous reality of a real-world encounter. The listener is overcome by layered sounds, music, and poetry, leaving an experience of the book that is composed of a variety of elements at once. The sentiment of the above poem is strikingly similar to the overall effect of the audio project with Grubbs. Drawing language from the legible words in the poems, readers are confronted with the text's commentary on hearing as an uncertain process and, consequently, on the audio project of "Frolic Architecture" as a product of this poem's self-reflection.

Howe's poem comments on the value of fractured poetic spaces and the haunting peace that comes with them. She begins with sound, claiming that when sounds or perhaps words are "heard as what might," they maintain "the mystery of the / original wild unbounded place." Reminiscent of the phrase "come what may," the opening lines invoke an acceptance of uncertainty and even a lack of comprehension. In such uncertainty, "place" is able to remain "original" and "unbounded." Although the speaker's use of the word "wild" may appear to have clear ecocritical connotations, it is used in a more general sense to indicate disorder and not to invoke nature specifically. Rather, "wild" here gestures toward the "somewhat broken" experience that emerges when one allows sound and, indeed, language to be pure ("heard as what might"). In such acceptance, the speaker reflects that "rest" or acceptance comes in the "empty / absolute" actually being seen as "infinite." The "broken" or fractured experience that the speaker proposes leads to an understanding of such chaos as an appreciation of the countless perspectives that can shape how one experiences a moment, whether textual or material.

The poem does more than create fractured space, though. It also grants text agential power in shaping the reader's experience of the poem and, further, the reader's mediated encounter with the Edwards archive in the Beinecke. As Howe adapts sections from the papers into the poems, she places her reader in contact with the original papers, but she does so through fragmentation and extractive poetics that recontextualize the original documents. Still, the reader's experience of the archives is forever changed, altered not only by the words on Howe's pages but also by the words and textual fragments themselves. Put another way, the poet's language acts beyond her control. As she includes incomplete scraps of text, bits of words, and fragments of letters, the text attains a power of its own. In its appearance on the page, it becomes a thing in itself – not a referent to the world, but an object. The partial letter that ends the first line of the most prominent section of the previous poem, for instance, evades the reader. It may be the letter *h,* but such a conclusive reading is impossible. Instead, the mark remains in the text, but it does not work toward meaning. Nevertheless, the mark is a shaping force in the reader's experience of Howe's poem and, as a result, the reader's interaction with the Edwards archive. The incompleteness of the mark gives it power; the reader may ascribe it meaning by deducing its referent, or it could be ignored entirely. Either way, the text remains dominant because ultimately it does not refer and it is not absent. Any deviation from the illegible mark is a departure from the text, and thus an alteration of its meaning. Ultimately, the uncertainty of the mark is both powerful and meaningful not only because it changes the meaning of Howe's poem, but also because it shapes how readers experience the poetic space of *That This* and the secondary space of the Beinecke.

To some extent, such a reading of the poem depends on Howe's subsequent collaboration with Grubbs, because the audio project provides a spatial context for the poems in the book. Through layering, the audio project highlights multiplicity and demands the acknowledgement of divergent perspectives. When read together, then, Howe's collaboration on the audio project and the original book, *That This*, are mutually clarified.

The third and title section of Howe's book reiterates the multiplicity and layering that is prominent in both the second section and the audio project. In section three, the poet returns to traditional poetic lines, engaging in self-reflexive commentary on the nature of experience and language's ability to capture it. The fractured reality of experience that is highlighted in the first two sections is examined in the final portion of the book. One poem reads:

> Is one mind put into another
> in us unknown to ourselves
> by going about among trees
> and fields in moonlight or in
> a garden to ease distance to
> fetch home spiritual things (104)

Although the speaker begins by establishing a splitting of the mind, she explains that such a split occurs "by going about among trees" and entering other natural spaces where she might try to "ease distance" between herself and nature and locate a "spiritual" center in the world. It is in the quest for one-ness with the physical world, then, that the speaker discovers the inability to ever be one with nature. She reflects that it is "by" engaging in such a search that her split mind emerges. The poem provides context for some of the book's earlier sections, revealing that the quest for a home or spiritual center instead produces fragmentation, splitting, and uncertainty. As the book demonstrates, such belonging comes not through single voices, clear meanings, and direct transcription, but instead from fractured spaces, where multiple perspectives, various senses, and necessary uncertainty are preserved.

Conclusion

In a 2012 interview in *The Paris Review,* Howe reflects on the spatiality of her poetics: "I often think of the space of a page as a stage, with words, letters, syllable characters moving across. In a certain way you can also say the poems in *Frolic* might be some sort of drawing" (qtd. in McLane). The poet's roots in visual arts and her continual push toward multisensory poetic environments lends her work to an ecopoetic reading. Recognizing the boundaries of language and the limitations of poetic form, Howe integrates sound, visual art, fragmentation, and prose into her poetry to make it a multisensory encounter. She emphasizes that experiences are shaped by multiple elements and differing perspectives, a reality that her work actively attempts to capture. As her comment indicates, her poems are active forces, in which text functions like a character in itself—powerful, active, and material.

That This also serves the added function of shaping the reader's experience of the world beyond the poem. Through her engagement with the Edwards family papers at the Beinecke, Howe's text influences how individual readers, most of whom have not visited the library and likely never will, perceive and interact with the archives. Her poetry uses uncertainty, fragmentation, and appropriation to highlight, first, how the poet's experience of the papers shaped her lived experience and poetic rendering of her husband's death, and second, how the reader's encounter with the text shapes his or her engagement with the distant archive. Through Howe's collage, what is to most readers an imagined space becomes real and active. Text, in its various iterations, is active for Howe, and that agential power allows her poems to put readers into the fractured spaces of real-world encounter.

Endnotes

[1]   Knickerbocker discusses this in his introduction of the text, where he recounts the book's inspiration from "the colonialist Reverend Hope Atherton's account of his actual wanderings in the New England wilderness in 1676" (173).

[2]   In her interview in Keller, the poet recalls her foray into creating artists' books, where she recalls creating books that blended words and images by using "flash association between the words and the picture or charts" beside them. She also comments on the development of her formal methodology when she recounts how she "always left a lot of white space on the page" (5).

Works Cited

Bass, Chloë. "An Architecture of Poetry and Sound." *Hyperallergic*. 11 Dec. 2013. https://hyperallergic.com/97949/an-architecture-of-poetry-and-sound. Accessed 1 Sep. 2015.

Bloomfield, Mandy. "Palimtextual Tracts: Susan Howe's Rearticulation of Place." *Contemporary Literature*, vol. 55, no. 4, 2014, pp. 665-700.

Grubbs, David, and Susan Howe. "Frolic Architecture." 1 Nov. 2011, Harvard University, Cambridge, MA. Performance.

Haraway, Donna. *How like a Leaf: An Interview with Thyrza Nichols Goodeye*. 1998. Routledge, 2000.

Howe, Susan. *That This*. New Directions, 2010.

Iovino, Serenella, and Serpil Oppermann. "Theorizing Material Ecocriticism: A Diptych." *ISLE*, vol. 19, no. 3, 2012, pp. 448-75.

Keller, Lynn. "An Interview with Susan Howe." *Contemporary Literature*, vol. 36, no. 1, 1995, pp. 1-34.

Knickerbocker, Scott. *Ecopoetics: The Language of Nature, the Nature of Language.* U of Massachusetts P, 2012.

McLane, Maureen N. "Susan Howe, The Art of Poetry No. 97." *Paris Review,* vol. 203, 2012.

Naylor, Paul. *Poetic Investigations: Singing the Holes in History.* Northwestern UP, 1999.

Biopoetics: The Interfaces between Language, Form and Life

MARGARET ANNE CLARKE

> *"What I cannot create, I do not understand"*
> *—Richard Feynman*

As this millennium continues, it is becoming ever clearer that amid the accelerating changes that are taking place in the fields of science, technology, and the arts, an increasing tendency is evident toward a new connectivity and a new rapprochement between areas of human knowing and discourse, hitherto held to be separate and compartmentalised. These conceptual transformations have profound implications for the poet, the poetic craft, and the nature of poetry itself. There is an increasing acknowledgment by contemporary poets and thinkers of the role they may fulfill in challenging, but also engaging with, the scientific paradigms that have, in large measure, shaped contemporary thinking concerning the nature of the world and human civilisation. Both the natural sciences and the poetic arts are engaged in the quest for knowledge and insight concerning the human condition and the environment that has shaped that condition. More specifically, both fields of knowledge have been concerned with healing the fracture between the disparate modes of knowing caused by diverging conceptions and uses of language. The quest of the poet to recover the ancient prophetic powers of the poetic arts to overcome and to integrate the fractured domains of existence has been a consistent theme since the Romantic era. Yet this quest has assumed a greater urgency in the current age, as science continues to reveal with increasing precision the linguistic foundations that underpin life itself. Thus the subject of this essay is the recovery of these linguistic foundations for the purposes of poetic creation through the theory and practice of 'biopoetics'. This genre was first defined and realised by the Brazilian-American artist and multimedia production practitioner Eduardo Kac, born in 1962 in Rio de Janeiro, Brazil.[1]

One overarching purpose has informed Eduardo Kac's long and prolific career as a multimedia artist, philosopher, and academic: a consistent and relentless quest to stretch the possibilities of his aesthetic practices beyond standard definitions and accepted uses. Yet this quest has also been realised through a process of continuous dialogue and engagement with the artistic and philosophical movements of Kac's era. Thus Kac has contributed on a global scale to contemporary debate and reflection on the nature of knowledge and aesthetics; the function of creative forms and their purpose, especially their public purpose; and the role of the artist and poet (Baldwin). In particular, the full gamut of Kac's work has integrated philosophical considerations surrounding synthetic biology, biotechnology, and genetic engineering, which is the conscious manipulation of genetic material, including 'cloning' and cross-fertilisation of species for a number of purposes. This concern has been an informing theme, indeed the most publicly known, of Kac's work in different types of media and other contexts. As an artist and practitioner, he is one of the best-known exponents of the genre of bio-art and biotech art, which, according to Gessert, has played a significant role in engaging the public with issues related to the human body, the nature of life itself, and its manipulation by human society (145).[2]

Biopoetics, as defined by Kac, represents a distinctive genre that falls within these broad paradigms. It is specifically defined in Kac's manifesto, which summarises biopoetics as 'the use of biotechnology and living organisms in poetry as a new realm of verbal, paraverbal and nonverbal creation' ("Biopoetry"). Kac then lists twenty examples of works that could be created with material in vivo, integrating these broad semiotic considerations. He also illustrates the principles of biopoetics more specifically with two works already created and publicly exhibited. The first, *Erratum I* (2006), is an exemplification of the poem as ecosystem, 'a living work that changes in response to internal metabolism and environmental conditions' ("Biopoetry"). Secondly, the installation *Genesis* (1999-2013), which Kac termed a 'bioproduction' (Medeiros 10), is a work created with proteins and amino acids that have been 'written' by Kac himself. This composition effects the mutating reproduction of a verse from the biblical book of Genesis, encoded into DNA (deoxyribonucleic acid, the genetic material within each living cell).

Both works explicitly engage with a range of issues in the field of biological science, encompassing the origin, evolution, and interrelationships of living organisms and their genetic codes. These are a means of authorship for Kac, who, in both pieces, integrates the semiotic properties of these organisms with human-created linguistic systems to create an interface combining the material and organic, within which 'language, form and life intersect' ("Biopoetry"). But the pieces are also consistent with Kac's purpose in engaging with what is at stake, socially, politically, and ethically, in the recently acquired powers over the natural world that biotechnology and genetic engineering have enabled. The aim of this essay, then, is to elucidate and reflect on the principles of biopoetics with the basic definitions given by Kac as my point of departure. After tracing the evolution of Kac's trajectory in the context of theoretical considerations on the nature of poetic creation from the mid-twentieth century and beyond, I then consider in more detail Kac's assertion that the 'paraverbal and nonverbal' is a fundamental component of the poetic form. To illustrate, I provide a closer reading of *Erratum I* and *Genesis,* which demonstrate in more detail the means and methods of biopoetics. The essay ends with broad conclusions drawn concerning the nature of language itself, its function, and its true origins.

Toward a Definition of Biopoetics I: Fracturing Away from Modernism

As we have seen, Kac's international standing over the past two decades has rested on works composed in numerous configurations. Because his works are often public installations, they have frequently been classified as works of art, and Kac himself is known, especially in the popular press, as an 'artist' (Pallardy; Frank Rose). Yet Kac's works, and the works of his colleagues, defy classification into any one specific genre, combining as they do living organisms, multimedia, robotics, telecommunications, and performative methods of all kinds, not infrequently involving the artists' own bodies. The problematic issue of classifying these productions—and, indeed, all artefacts of 'digital' or 'new media' that are not confined to any one typographic or artistic medium—into specific genres has been pointed out by theorists such as N. Katherine Hayles and Espen Aarseth (Hayles, "The Time" 181-83; Aarseth, *Genre* 46 – 49). What is it, then, about Kac's defined poetic genres, forms, and methods that distinguish them from art, or from multimedia installations and exhibits commonly defined as art?

One basic distinction between art and poetry is that art is held to be presentational or representational. Art embodies in its aesthetic form content that is affective, while poetry may share some of the characteristics of art; poems of all genres are a composite of different elements producing diverse effects on the reader. Poetry is nonetheless still discursive in nature, always involving a vocabulary, syntactical formation, and a determinate meaning – a linguistic construct, in other words (Langer). Nonetheless, as Deac also points out, Kac has, in fact, persistently emphasised the fundamentally discursive nature of his pieces in this sense and frequently referred to his productions as 'poetry' in his copious theoretical writings, manifestos, and interviews. Furthermore, he has classified these very precisely in generic formats. Another dedicated page on Kac's website lists four genres of poetry that have been created and defined: holopoetry, digital poetry, biopoetry, and space poetry ("Media Poetry"). Another, more recently created genre, aromapoetry, was the subject of Kac's 2013 exhibitions (Brown). Each genre is created through a media not hitherto considered as a means of poetic creation. For example, the genre of holopoetry consists of poems created through the physical properties of energy and illumination using remote teleprojectors. Space poetry entails, according to Kac, the creation of 'a new poetic language that participates in the creation of the new space culture through the exploration of the creative potential of low gravity and weightlessness' ("Space Poetry"). Each definition represents an extension of poetry into different spaces; furthermore, as these examples illustrate, each genre invented is defined by the medium used to create it (Deac).

For all the innovation inherent in these works, Kac still might not seem to be attempting something that radically changes the nature of poetry; the spatial dimension is, after all, inherent in the poetic, as opposed to the prose form, and is frequently isomorphic with material shapes and artefacts. Thus the purpose of poetic form, according to Robin Allott, is

to provide a perceptual vessel for 'the living stuff of the imagination' (Allott 84). A poem is a structure that is designed to
preserve words and ensure their effect and lasting persistence
in the consciousness of the reader. Words and their syntactic
arrangements are preserved within the poem's structure, and
the interaction of each contributes to 'the multidimension-
al structure' in which conceptual meaning is generated and
regenerated through the consciousness of the reader or the
viewer. The poem, then, 'seems to operate as a unity of all its
powers' (Allott 85). But in order to operate as a true unity, the
poet must also find some way of transcending the perceived
one-dimensional status of words as signs on a page or refer-
ents to some concept outside their existence, revealing the
processes by which they come into being in relation to the
material essence that 'powers' their existence and generation.
This process enables the reader of the poem to seek out and
engage with the meaning-making properties between the figurative and the material, often represented and embodied in
the material itself.

This precise concern, and the resultant striving toward re-
alising ever more intricate connections among the material,
the visual, and the textual in order to realise a poetics greater
than, and indeed radically different to, the sum of its parts, has
been the defining characteristic of Kac's work since the beginning of his career in the early 1980s. The origins of Kac's poetics can be traced to early formative experiences in the intensive spirit of poetic and artistic experimentation within late
Brazilian modernism, in particular, the genre of concretism,
which took a diversity of forms as an international movement

from the late 1950s. Yet concretism precisely attempted to re-frame modern poetics in terms of the key word verbivocovisu-al – that is, the status of the word not only as sign but also as image, as well as phonetic and sonoric properties of poetry, and the possibilities inherent in its materiality.

Kac's aims in the early years of his trajectory, therefore, still amounted fundamentally to an expansion of these char-acteristics of poetry and poetics developed in the mid-twen-tieth century. The reflexive nature of modernist poetics also entailed a close consideration of how a poem was written and therefore led to a renewed attention to the material of which it was constructed. In Clement Greenberg's often-quot-ed words, 'The unique and proper area of competence of each art coincided with all that was unique in the nature of its medium' (51). The practitioners of modernism consistently, and in many different modes, sought to recover the word 'as a unity of all its powers': the correspondences between sound, colour, the material, and typographic were exploited and de-veloped within various themes and manifested in many dif-ferent forms.

From the mid-1960s onward, however, the broader trends of modernism deepened and accelerated still further, as the recombinant works integrating the visual cultures of image, film, and sound were remediated through electronic net-works, linked together and made explicit on a computer in-terface. Writing and textual practice that engaged, and en-gages with, digital, programmable media and operating in a networked context made ever and progressively clearer the *techne* and process by which a poem, and the word, was brought into being. Digital and 'new media' poetry, therefore, attempted to evoke a network of material-semiotic relations (Livingston 104) and enabled the creation of numerous combinations between the linguistic properties of the poem and the media within which they were inscribed (Glazier 1-3). The new media poetry recovered its nature as a phenomenon that was relational and dialogic in many senses. Moreover, the networks through which the poetry was disseminated enabled a more public praxis and a more substantive engagement with the political and social themes of the age, more consonant with a global mass media society.

The reading and reception of these works in a public sphere, exhibited in public spaces, and networked globally over the internet also enabled a new relation with the reader of the work as an 'ergodic' participant (Aarseth, *Cybertext* 2-5) and co-creator of the poetic piece. Equally, as the material properties of art were progressively accentuated, poems were composed in such a way as to emphasise the performative aspect of words. As Livingston points out, performance has always been an inherent characteristic of the poetic form: 'Poetry, even in a relatively narrow sense, is always a performance and negotiation of power relations in language' (95). The language of the new media poetry increasingly emphasised performance and action in language, together with the poets' exposition of the process by which they created and what Noble (137) calls the 'doingness' of things, the processes that occur, rather than the poet as subject who is the possessor of the being-ness or doing-ness. All these features are also characteristic of Kac's poetic genres, and as we shall see, they were foreshadowed and developed in *Erratum I* and *Genesis,* particularly the latter.

Towards Definition of Biopoetics II: Fracturing Away from Even 'New' Media

Kac's subsequent work, frequently in tandem with his colleagues, has been dedicated to developing these possibilities to their full potential, as the digital and technological affordances available to him and his contemporaries became more sophisticated from the late 1990s to date. Yet as clearly demonstrated in his writings and interviews, a fundamental restlessness and desire to transcend even these ever-shifting and evolving paradigms has also been a characteristic of his work. I would argue that this is due to Kac's sense, often articulated in his interviews and theoretical works, that digital and new media poetry continued to maintain the basic schism between the poetic word and the inanimate matter and energy within which it was inscribed ("On the Notion" 21-22). In that sense, late twentieth-century poetics, for all its experimentation, maintained its status as an outpost of late modernity.

As we have seen, contemporary avant-garde poetic movements in which Kac has played a key role have formulated many new and varied genres of poetry around the particular combinations between language and the medium within which it is inscribed; they have opened the way for a radical rethinking of the role of poetry but have not challenged one

fundamental assumption: that concepts of 'life' in all paradigms of poetic creation are exclusively located in the being
of the poet as living subject. Any transformative power that
a poetic text may have resides in its status as a product of
human agency. The divide between human culture and organic nature, and the domination of the former over the latter, has
not been fully confronted or overcome. If this is the case, then
the relationship between the material and the symbolic and
figurative, or between technology and human subjectivity, remains a heavily contested situation.

Paradoxically, this divide became attenuated as the digital
era progressed. In the virtual realm of the 1990s and the beginning of this millennium, knowledge, now coded as information or data, was circulated in an unmodified form across
and through different forms of materiality or, as N. Katherine
Hayles puts it, between 'carbon-based organic components
and silicon-based electronic components' (*How We Became*
2). Theories of the 'posthuman', as propounded by Hayles and
Rosi Braidotti, no longer exclusively located the human subject within one individual body or consciousness, but rather
they distributed these components across biological and
technological systems (Hayles, *How We Became* 2-7; Braidotti
3-12). Yet theories of 'distributed cognition' still did not de-
finitively succeed in addressing fundamental questions concerning the human subject's relations with the biological and
inanimate spheres. For example, the numerous hybrid forms
between the living and the inanimate machine conceptual-
ised by the generic term of the 'cyborg' or any other networks
of cross-wired social and natural phenomena still maintained

these conceptual opposites (*How We Became* 6-7). In the field of poetics also, the posthuman paradigm still tended to privilege informational pattern over material reality: 'Out of habit, we identify the "modernist" poetic text as "materialised" and the "postmodern" poetic text as "dematerialised" and ephemeral' (Lennon 65). In this sense, as Rabinowitz and Geil put it, 'Digital culture needs to be understood as at once an outpost of Enlightenment thought and an agent of its steady erosion' (Rabinowitz and Geil 7).

Kac's works, therefore, have consistently demonstrated a consistent striving toward attaining a final and complete integration of the word and its medium. In common with his contemporaries, he has attempted to affect this through combinations or conglomerations of media and has thus contributed toward the concept of the poem as embodying 'the transition from the one-medium work of art to the inter-medial and interlingual conglomerate' (Deac). In 1995, for example, Kac stated, 'With computer holopoems I hope to extend the solubility of the sign to the verbal particles of the written language... widening the gamut of rhythms and significance' (48). Kac also expanded on the same principle in a published interview with Annick Bureaud in 2007, stating:

> I'm interested in what could be described as a semiological continuum between making a line that you recognise as a letter and making a line that will have a form that does not carry meaning verbally but through other modalities of representation such as an icon or symbol ... not to make the transition from one to the other but to present to the viewer the very experience of uninterrupted transition.

Uniting Fractured Epistemologies and Biopoetics as Ethical Practice

This informing aim of Kac's, which is to realise the experience of 'uninterrupted transition' between the word and its material environment, is also intimately connected with ethical concerns concerning the destruction and harm to species other than the human. Language is profoundly implicated in this issue: as a function of a human subject and mind held to be uniquely evolved above every other entity on earth, language is the instrument that has enabled human beings to gain dominance over nature, frequently to its harm or destruction. Many contemporary theories and current conceptualisations concerning the nature of language derive from the inheritance of Newtonian scientific philosophies of the sixteenth and seventeenth centuries, together with dualistic Cartesian ideas concerning the self-directed autonomy of the human mind, independent of the senses and detached from the 'natural' world.

This inheritance, together with the ever more specialised functions that the Industrial Revolution and ensuing divisions of society and labour entailed, had lasting implications for the way language is defined and used (Midgley 2). The separation between and within linguistic systems and their components is inherent to the specialised divisions between language and the material environment that the intellectual capacities of humankind must necessarily, according to scientific thought, manipulate. For both the reader and the user, language and medium are operationally different kinds of unities that generate nonintersecting phenomenological domains. This inheritance of certain 'scientific' modes of thinking has given rise to the convention that science and poetry are diametric opposites, especially in their 'modes of knowing' (Saunders 6), which refers to ways of conceptualising reality, life, and consciousness. Thus many scholars have pointed out the difficulties in bringing about rapprochement and understanding, let alone finding common ground, between two such seemingly different forms of knowledge (Holmes, Levine, Middleton). Peter Middleton, for example, emphasises the lack of common agreement and 'seemingly impenetrable epistemic barriers between science and literature' (40).

What, then, is Kac's solution? Beyond overcoming the division between the sign and its materiality, Kac aims to extend our common concepts of language and poetry beyond the textual and written inscription, to also encompass nonverbal semiotic systems and the process of communication in its entirety. In another interview, Kac stated, 'To summarise language, the semiological continuum between sign systems, communication, subjectivity, the organic continuum between every kind of life, and the dialogic are the paradigm of relationships that have shaped my interests from the very beginning' (Kalenberg).

Communication systems may be transmitted by electromagnetic, chemical, or physical means; thus language, in this expanded sense, is an inherent part of the medium from which it evolves and, through complex processes, coalesces into a coherent and self-sustaining structure (Stenlund 17-22).

Again, this is not a unique concept: within the broad theoretical concept of cybernetics as applied to fields such as computing, sociology, and biology, phenomena defined by their communicative, relational, and self-regulating systems have constituted another broad conceptual shift in post–World War II society. As applied to poetry and poetics, these concepts have yet to be fully developed (Schleifer 1-22).

Kac's solutions lie partly in the changing concepts of science itself, and especially in the insights afforded by the biological sciences: what John Barnie terms 'the border between the new biology and the new humanities' (55). According to the 'new biology', language and the ecologies of human communication are also dynamic systems and consist of a network of components that have emerged, ultimately, from the biochemical networks that create life (Livingston 104). This concept has been made ever more explicit with the acknowledgment that life itself originates from a defined language, or code, which is generative in nature and, indeed, has been defined as a 'grammar of life' (Middleton 50). In fact, the nature of evolution and the generation and regeneration of all forms of life are now commonly accepted as linguistic in their origins: a process of transmitting information through chromosomes, structures inside the nucleus of each living cell and composed of the molecule DNA. This contains the code – pairings of chemical letters – which, according to Adam Rutherford, is 'an instruction manual for all the processes of life, including the very instructions required for the replication process itself" (33-34).[4] According to Michael Denton, each cell with genetic information consists of artificial languages and their decoding systems, memory banks for information storage and retrieval, elegant control systems regulating the automated assembly of parts and components, error fail-safe and proof-reading devices utilised for quality control, assembly processes involving the principle of prefabrication and modular construction... [and a] capacity not equalled in any of our most advanced machines, for it would be capable of replicating its entire structure within a matter of a few hours. (329)

In Kac's poetry, the biochemical actions from which the 'living stuff of the imagination' evolves are made manifest, and the processes through which the poetic form finally results are made fully apparent. Further, in much of his work, there is an explicit convergence with the information systems that underpin both biological life-forms and genetics. The third subgenre within the field of biopoetics illustrates Kac's expansion into the field of biotechnology, the conscious manipulation of the codes from which living organisms evolve. In an interview with Lisa Lynch, Kac stated:

Biotechnology, as the structuralists were keen to point out, can be compared to a language.... Like any other technology, biotechnology operates through the material semiotics that Donna Haraway writes about so eloquently. There is a process of sign operation that is not verbal or visual, that is very much about changing things in the world.

But the concept of life as resembling an orchestra, a self-composing pattern and its involvement in the 'expression of high-level function' (Noble 24), carries the possibility for an expanded role and radical possibilities for the artist and poet. The role of the poet is not only to compose with words, but to be a creator of the material conditions within which these words are embedded, and on which they depend for their sustainability as a closed system with permeable boundaries. Thus the poet creates a 'language of life' that is self-perpetuating and sustainable. Kac has integrated into his poetics the full range and possibilities of those communication systems, transmitted by living species other than the human and through means other than the typographic; these systems also encompass the informational codes within living organisms that generate and evolve life itself. This principle is illustrated clearly on Kac's dedicated page ("Biopoetry").

After setting out his fundamental method, consisting of 'the use of biotechnology and living organisms in poetry as a new realm of verbal, paraverbal and nonverbal creation', Kac then proceeds to exemplify with a list of twenty types of works that could potentially be created under his paradigm. Three principles or methods are evident in these creations: Firstly, the integration of communication mechanisms evolved by species other than *Homo sapiens* – the dances of bees, the sound waves emitted by marine mammals, the songs of avian species such as parrots. Secondly, the creation of artefacts using varied organic means – seeds, agar, culture tissue – in which the poet inscribes written signs that evolve or disintegrate according to the resulting organism's growth. Thirdly, the overt manipulation of molecular life, involving also the use of proteins and amino acids in works such as amoebal scripting and proteopoetics, which assume the character of laboratory experiments as well as the creation of artefacts for aesthetic purposes.

Erratum I: The Principles of Material Semiotics, and Poem as Ecosystem

Erratum I is an exemplar of one of the twenty poetic forms proposed on Kac's dedicated page ("Biopoetry"). The work is defined as a 'metabolic metaphor'. The method of composition, as described by Kac, is to

> control the metabolism of some microorganisms within a large population in thick media so that ephemeral words can be produced by their reactions to specific environmental conditions, such as their exposure to light. Allow these living words to dissipate themselves naturally. The temporal duration of this process of dissipation should be controlled so as to be an intrinsic element of the meaning of the poem. ("Biopoetry")

Erratum I is a poem constructed as ecosystem—that is, a biological community of organisms interacting with their physical environment. Kac calls this poetic genre a 'biotope', a living work that evolves in response to an internal metabolism and to the dynamics of the environmental conditions in which it is exhibited, including temperature, relative humidity, flow of air, and levels of light. Within the biotope lies a verbal composition embedded within a composite of inorganic and organic matter, made up of microscopic living entities and media consisting of natural elements: earth, water, and other materials. A continuous sequence of images can be discerned, illustrating pairs of words such as 'Wind/Mind'. Layers of colours form at the exact moment when the verbal forms begin to dissolve. In this way, the boundaries between the materiality of the poem and its semiotic and discursive structures exist in a process of continuous interchange and mutual dependency for their survival and regeneration. The ecosystem depends for its survival and self-sustenance on both internal and external factors: the living organisms must be sustained in conjunction with the inanimate components of their environment—air, water, and mineral soil.

Kac's adaptation of 'material semiotics' consists of constructing a material environment that is propitious for words as organisms to evolve and replicate. In doing so, he illustrates what James Lovelock terms 'the hierarchy of intensity' between organic and inorganic matters (*The Ages* 42). *Erratum I,* then, also engages with the increasingly contentious issue of defining life. If definitions of life are expanded or disputed, especially in the domain of the animate or the inanimate, this has implications for the material with which the poet has to work. Mary Midgley (283-84) points out that the concept of life applies to entities operating on different scales. The key implication, however, is that life consists of a continuum between the animate and the inanimate. This is also a core premise of Lovelock's *The Ages of Gaia,* which concludes by denying any clear distinction between living and nonliving material. According to Midgeley: 'The crucial point is that life is not an accident or an alien invader but something which has grown out of the earth itself. The sharp divisions we make across this continuum reflect academic specialisations rather than unbreakable natural barriers' (204). In that sense, *Erratum I* is an exposition of the philosophy of Gaia: that the nature of life itself, and the poetic construct it generates, is a self-regulating, self-sustaining principle. A state of mutual cooperation and symbiosis is the precondition for both creation and innovation in evolution, implicated also in the creative and, of course, the poetic process.

One important aspect of biopoetics, therefore, entails constructing a poem by creating the conditions within which symbiosis and feedback may take place, and the self-generation of the work as a closed yet permeable system may be sustained. *Erratum I* also integrates the concept of creation as 'autopolesis', first introduced and expounded by Humberto Maturana and Francisco Varela (Vandestraeten 377), and which has been named as a key contribution to the 'conceptual revolution', gradually effecting a rapprochement and conceptual integration between the humanities and the biological sciences. According to Francisco Varela, 'The notion of autopoiesis is at the core of a shift in perspective about biological phenomena: it expresses that the mechanisms of self-production are the key to understanding both the diversity and the uniqueness of the living.' (Varela 14). More specifically, any living organism is defined first and foremost as a network of components that, through a process of continuous interaction and transformation, precisely gives rise to these mechanisms and maintains this process as a concrete, self-contained unity in a state of equilibrium.

The methods that Kac has adapted from this concept of life, however, are not unique to his biopoetics: his compositions have consistently aimed to achieve stabilising feedback (Lenton 445)—that is, the modification or control of a process or system by its results or effects, which enables poetry to sustain itself as a functioning material and linguistic system. In the case of *Erratum I*, this feedback takes place within a biochemical structure, and the principle of composition is based on the structure of all living systems: closure is realised through a continuous structural change in conditions of continuous material interchange with the medium, and this is the means by which autopoiesis can be maintained. Autopoiesis is also a communicative mechanism that unites the spheres of the animate and nonanimate: feedback is essential to the working and survival of all regulatory mechanisms found throughout living and nonliving nature. The informational feedback loop does not connect a system to its environment; nonetheless, the emphasis now is on the mutually constitutive interactions between the components of a system rather than the 'pure' message, signal, or information (Hayles, Foreword x-xi). This is what maintains *Erratum I* as a dynamic ecosystem in a state of continuous self-reproduction: an autopoietic system. To the extent that an autopoietic system is defined as a unity by its autopoeisis, its continuing function as an organic system is dependent on the fact that all its trajectories must lead to autopoiesis. If they do not, disintegration of the system quickly follows (Maturana and Varela 21).

Yet this raises another question: If a living organism is defined by its capacity to reproduce, then how exactly does it reproduce, and in what sense are words 'alive'? Is the poet's role to reproduce words as living organisms or types of code that trigger life processes? The poet, in this context, adopts the role of what Maturana calls a 'super-observer' (56-59), regulating the system from the outside, in much the same way as a biologist would. Thus the poet's role, if he or she is working with these materials, is to act as the regulator, or orchestrate the living organism to be a self-sustaining structure and attain the homeostasis that will maintain the organism in its living state.[5] The poet must keep the system alive and further enrich the communication system created with inscribed words. As described by Loet Leydesdorf, the multiple components within the system continue to "select upon each other, thereby producing new recombinations that may be innovative... If codification succeeds, a new pathway for propelling the communication is generated" (76). The creation of the poetic structure is therefore a coevolutionary process: words and medium shape each other mutually along these trajectories by selecting upon each other in terms of signals and noise, which become recursive as the new recombinations are repeated and then further recodified.

Yet these codes, inscribed in genetic material—the originators and transmitters of language – may operate only as 'signs' insofar as they have the material conditions to do so, and they themselves also act on their material circumstances. Thus an essential component of poetic creation must entail the exploration and full integration of a sign – however constructed – with its material and organic environment. The emphasis in this field of biology, then, is on integrative processes and systems: things have a relational aspect to one another, not an atomistic one, and this is directly counterposed to ideas of genetic determinism and genetic reductionism. It has been argued by biologists such as Steven Rose that mutual coop- eration and symbiosis, not competition, is the source of inno- vation in evolutionary systems. He asserts the fundamental characteristic of the gene as 'an active participant in the cel- lular orchestra' and that 'the organism is both the weaver and the pattern it weaves, the choreographer and the dancer that is danced' (125-6, 171). This idea, in its most influential form, has also been expounded in Lovelock's *Gaia,* and is the informing concept behind the second work chosen by Kac to exemplify the practice of biopoetics, *Genesis.*

Genesis: The Ethical and Epistemological Func- tion of Biopoetics, and the Creation of Commu- nicative Communality

The installation *Genesis* further engages with what Adam Rutherford terms 'the bedrock of biology: the translation of code into action' (37) and represents another poetic appropri- ation by Eduardo Kac of intricate networks of chemical reac- tions that drive the principles of life and, ultimately, language. If science has now enabled new generative properties, so may language also assume a generative function and emphasise the primacy of the code, or signifying, in the creation of life. One major theme of Kac's work, albeit a controversial and contentious one, derives from the principle that DNA code is a language that remains precisely the same across all forms of life: a universal language with a common origin. A single-cell bacterium gives instructions of precisely the same order as those of a human organism. Kac's work thus engages with what is known as 'the singularity', defined in the following way by Adam Rutherford: 'Those three aspects of biology – cells only from existing cells, DNA changing from imperfect copying, and modified descent of a species as a result – logically unveil a single line of ancestry that inevitably leads back to a single point in our deep, deep past... inevitably back to the origin of life' (42). This concept therefore must also entail the quest for the origins of human consciousness and language, not, in fact, a result of humankind's 'higher intelligence' or, indeed, of human agency at all. These origins reside in the very structure of life and mutate into human-created linguistic systems by a process of 'translation' across organic and inorganic systems, exemplified by Kac in *Genesis.*

As the installation's title implies, Kac integrated the principle of the singularity with the Judeo-Christian creation myth of Genesis, cognate with 'gene', which, in common with other creation myths such as the *Popul Vuh,* conceives of creation in both scientific and mythical terms: as an infusion of energy in the form of 'spirit', a division between the elements, and a basic patterning in order to create the conditions for life, and therefore language, to emerge. The composition of *Genesis* exemplifies Kac's statement in an interview with Annick Bureaud in 2007:

> The text 'Biopoetry' stated from my recognition, which was realised in the work *Genesis*, that the living can be a writing medium. With *Genesis* it is not only the inscription of meaning, but the inscription, the processing that takes place inside the living and then the output. You could think of an expanded notion of writing that involved the living.

The process by which Kac composed *Genesis* imitates the alteration and remix and reassembling of basic genetic codes, a process that bears the generic term of 'synthetic biology', from which originated the term 'genetic engineering'. This procedure usually involves minor modifications to already existing organisms for stated, generally functional, purposes. The medium for these modifications is generally bacteria, or single-celled organisms, whose genetic processes are sufficiently well understood to adjust without difficulty, and which also constitute Kac's principal medium in *Genesis*. The work consists of a codified verse taken from the biblical book of Genesis in the King James Version: 'And God said, Let us make man in our image, after our likeness: and let them have dominion over the fish of the sea, and over the fowl of the air, and over the cattle, and over all the earth, and over every creeping thing that creepeth upon the earth' (Gen. 1.26). This passage was subsequently translated into Morse code and then into DNA code. With the help of geneticist Charles Strom, Kac had a biotech company synthesise the gene 'written' by the code. Kac termed this an 'artist's gene' (Kalenberg) and inserted it into bacteria culture, transferred to a Petri dish: this was the centrepiece of the installation.

To 'see' the genetic artwork, viewers could hit a toggle button, which activated a UV light trained over the Petri dish. When activated, the UV light simultaneously illuminated the bacteria and mutated it, producing alterations in the synthetic gene through the ultraviolet exposure. On each occasion *Genesis* has been exhibited in cities in Brazil, Europe, and the United States, both a new 'translation' of the Bible verse and a new strain of bacteria have been produced. When the artist reversed the process, translating the genes back into Morse code and then to English, the verse emerged considerably modified from the original.

The creation of a poetic construction such as *Genesis* involves a fundamental challenge to the epistemic functions of poetry and the further extension of its possibilities as a generator and transmitter of knowledge. To begin with, Kac places greater emphasis on interrogating the semiotic and linguistic properties within the material universe that scientists investigate in an empirical fashion; this also relates to the expansion of the definition of language to embrace other semiotic and communicative systems, which I have traced within his trajectory as a whole. The process of 'inscription' named by Kac is a communicative process that entails three primary steps: The first step is to identify the information that exists embedded in the medium. The second step entails the coding of the information in a comprehensible form by the poet to a receiver. Finally, the receiver, in this case the viewers or audiences for

Kac's work, translate the information into a concept that they are able to comprehend.

Thus both the poet and the poetry itself assume a role as translator and communicator of the code from which life is held to originate. This role also implies other issues related to the relational and dialogic properties of *Genesis*, which are the third 'pillar' of biopoetics as defined by Kac. Since communication systems also involve the meaningful exchange of information between their participants and a message from the sender to the receivers, they require an area of 'communicative commonality' where the viewers and readers may also participate in the evolution of the poem. To the explicit engagement with biological systems that Kac enables with *Genesis,* he adds the participation and formation of a community from 'distributed communication' (Rabinowitz and Geil 5). Thus *Genesis* is composed in such a way as to enable the active participation and experimentation of the 'lay' public. All this assists in constructing what Kac, following Maturana, termed 'consensual domains: shared spheres of perception, cognition and agency, in which two or more sentient beings... can negotiate their experience dialogically' (Lynch).

Based on these precepts, the cognitive, epistemic functions of poetry are developed further, in the sense that poetry should be a vehicle and instrument for knowledge, disseminated in a public sphere. Thus *Genesis* illustrates Kac's disruption of the 'epistemic hierarchy' within which scientific knowledge and its manipulation are assumed to be solely the provenance of trained scientists and the corporate institutions within which they operate. One informing purpose of the installation, therefore, is to challenge the status of the natural sciences at the apex of the 'epistemic hierarchy' – that is, the analysis and rational judgment of empirically provable facts and material phenomena for the purpose of corporate or public benefit. This mode of knowledge occupies a status in society above the not easily evaluated subjective and hermeneutic nature of poetry. Middleton, however, In common with Kac, goes on to suggest that this schism, or fracture, might be at least partly breached by a diminishing of focus on the nature of knowledge and more on 'homologies of method' (qtd. in Holmes 11). Gillian Beer, indeed, suggests that the engagement of the humanities disciplines with science should not, in fact, have the intention of appropriating scientific method for aesthetic purposes, but it should serve as a means to destabilise accepted reasoning and preconceptions associated with these forms of knowledge (115).

This destabilisation has always been one of Kac's persistent and fundamental aims, given the wide scope of his progressive engagement with the biological sciences. Modern poetry may therefore serve as a mode of research and cognitive inquiry in a spirit of discovery that is more overtly creative in nature, producing an artefact that merges the aesthetic and the apprehension of knowledge. Therefore, metaphorically speaking at least, 'the aims and behaviour of poetry are those of research, and literally that the actual language used in the performance of the poem is the medium of its investigation' (Middleton 48).

By recovering the didactic and dialogic function the poetic arts used to serve before the Enlightenment era, Kac, among others, has been able to reappropriate the role of the poet as educator and griot. Yet this also implies that the role of the poet must assume a more public function, as the outlets for the dissemination of poetry become more diverse. *Genesis* is more accessible and overtly socially constructed, capable of public exhibition and diffusion through global networks, with more outlets for its dissemination than the printed book, and concerned with themes engaging with issues of global concern and social and ethical import.

Conclusion

In this essay, I have traced the development and characteristics of the theory and practice of biopoetics through two specific examples of the practice of artist, poet, and theorist Eduardo Kac. My point of departure is the idea that the evolution of the genre has necessitated a long and consistent progression through the polymorphous dimensions of poetic creation and the media by and through which the work is created. Eduardo Kac's poetic trajectory has been largely a riposte to, and an engagement with, the preoccupations of much poetic practice of the twentieth century. Modernist poetics, widening its scope through ever-expanding forms of media and electronic transmission, was concerned with desegregating the compartmentalised categories of language, in large part by assimilating these with their material composition. As the twentieth century progressed, this aim was further enabled by the relational and communicative potential of digital and electronic media, broadening the relations of time and space through which poetry could communicate. Thus, the defining purpose of Eduardo Kac's aesthetic throughout the sequence of phases described in this essay has been to seek an encounter, a reconciliation and integration among the material, semiotic, aesthetic, and linguistic domains, which have been fractured throughout post-Enlightenment modernity. While Kac's earlier work formed a necessary precondition for the development of his biopoetics, the diverse media he worked with in many different contexts and for many different purposes never finally overcame the schism between, on the one hand, the functional and figurative properties of language and, on the other, the inanimate matter and energy with which his poetry was composed and transmitted. Nor did Kac's creations effect a true metamorphosis beyond the parameters of creative agency, still exclusively located in the actions of the poet, or surmount the demarcations of human knowledge into categories that reflect the fractured nature of human experience and its conceptualisation of the world.

So one solution, posited in *Erratum I* and *Genesis,* was to relocate language itself beyond the consciousness and agency of the human through engagement with the evolving knowledge of the 'language of life', which the biological sciences have gradually revealed, and in the process, to expand and transform the role of the poet. The practice of biopoetics, which applies the code that generates living in vivo matter to create works that unite the material, discursive, and figurative, enables the fracture between the very different epistemological domains of science and poetry to be finally healed. By its very nature, the work entails a process that is still highly ex-perimental in nature and inconclusive. Nonetheless, *Erratum I* and *Genesis,* informed by current definitions from the bio-logical science concerning the nature of life itself, have estab-lished some principles by which this genre may be defined, as well as some methods and resources by which it may be creat-ed. These works, created with living media, envision and point toward a future direction for the role of both poetry and the creative process within which poet and audience, language and matter, scientific analysis and poetic insight could evolve and transform in a state of mutual cooperation and symbiosis. These creations therefore point toward a wider and more in-tegrating role for the poet, and poetic discourse itself, which, in the words of Novalis, 'heals the wounds inflicted by reason' (Corbett 30) in a world where questions and issues of the rela-tionship between humankind and the natural world assume an ever-increasing urgency.

Endnotes

[1] Kac's biography, together with comprehensive information about his works and full bibliography, is available on his dedicated website, www.ekac.org.

[2] Two projects in particular have garnered international publicity. Firstly, a transgenic art project titled 'GFP Bunny' was created in 2000 (GFP Bunny"). Its centrepiece was a living rabbit named Alba, bred from an egg inseminated with a green fluorescent protein of a Pacific Northwest jellyfish, which caused the animal's whiskers to glow green when placed under certain lighting. The project, intended by Kac to highlight the ethical concerns raised by animal experimentation and foster public dialogue, did indeed generate considerable controversy. Secondly, Kac and his collaborators created a "plantimal" as part of an exhibition at the Weismann Art Museum in Minneapolis in 2008. This, according to Kac, was a new kind of life-form called the Edunia, a genetically engineered flower bred partly from a gene isolated and sequenced from Kac's bloodstream and injected into a petunia.

[3] A complete illustrated anthology of Eduardo Kac's works defined as poetry, created from 1982 onward, was published under the title *Hodibis Potax* on the occasion of Kac's exhibition of the same name in the Biennial des Poètes, France.

[4] The iconic double helix of the DNA molecule is due to its composition as paired chemical letters: A for adenine, T for thymine, C for cytosine, and G for guanine. Each letter is precisely joined with a corresponding letter to form one rung of the helix, and each rung enables a copying mechanism for the genetic material. The runged letters are tightly wound up around beads of protein, and together these constitute a chromosome.

[5] Homeostasis is the property of a system in which variables are regulated so that internal conditions, and thus the internal environment, remain stable and relatively constant. Examples of homeostasis include the regulation of temperature and the balance between acidity and alkalinity.

Works Cited

Aarseth, Espen. *Cybertext: Perspectives on Ergodic Literature.* Johns Hopkins UP, 1997.

_____. "Genre Trouble: Narrativism and the Art of Simulation". *First Person: New Media as Story, Performance and Game,* edited by Noah Wardrip-Fruin and Pat Harrigan, MIT P, 2004, pp. 45-55.

Allott, Robin. "The pythagorean perspective: the arts and sociobiology." *Journal of Social and Evolutionary Systems,* vol. 17, no.1, 1994, pp. 71-90.

Baldwin, Sandy. "Art, Empire, Industry: The Importance of Eduardo Kac." *Electronic Book Review,* 5 Oct. 2007, https://electronicbookreview.com/essay/art-empire-industry-the-importance-of-eduardo-kac/.

Barnie, John. "The Poetics of Consilience: Edward O. Wilson and A. R. Ammons." *Science in Modern Poetry: New Directions,* edited by John Holmes, Liverpool UP, 2012, pp. 55-66.

Beer, Gillian. *Open Fields: Science in Cultural Encounter.* Clarendon, 1996.

The Bible. Authorized King James Version, edited by John Stirling, British and Foreign Bible Society, 1954.

Braidotti, Rosi. *The Posthuman.* Polity, 2013.

Brown, Alison. "A Rabbit by Any Other Name." *Kopenhagen Magasin,* 24 May 2013, http://kopenhagen.dk/magasin/magazine-single/article/a-rabbit-by-any-other-name/.

Bureaud, Annick. "Eduardo Kac: Living Poetry." *Art Press*, no. 332, Mar. 2007, www.annickbureaud.net/wp-content/uploads/2011/01/Kac InterEN.pdf.

Corbett, William. "Charles Simmic". *Poets and Writers Magazine*, vol. 24, no. 3, pp. 30-35

Deac, Ioana-Eliza. "New Meanings of Poetry in Eduardo Kac's Poems." *Cybertext Yearbook 2010*. http://cybertext.hum.jyu.fi/index.php?browsebook=7. Accessed 15 July 2017.

Denton, Michael. *Evolution: A Theory in Crisis*. Adler & Adler, 1996.

Gessert, George. *Green Light: Towards an Art of Evolution*. MIT P, 2012.

Greenberg, Clement. "Modernist Painting." *The Collected Essays and Criticism: Modernism with a Vengeance, 1975–1969*. U of Chicago P, 1995.

Glazier, Loss Pequeno. *Digital Poetics: The Making of E-Poetries*. U of Alabama P, 2012.

Hayles, N. Katherine. "Foreword." *Between Science and Literature: An Introduction to Autopoetics*, by Ira Livingston, U of Illinois P, 2012, pp. ix-xii.

————. *How We Became Posthuman: Virtual Bodies in Cybernetics, Literature and Informatics*. Chicago UP, 1999.

————. "The Time of Digital Poetry: From Object to Event." *New Media Poetics: Contexts, Technotexts, and Theories*, edited by Adelaide Morris and Thomas Swiss, MIT P, 2006, pp. 181-209.

Holmes, John. "Introduction." *Science in Modern Poetry: New Directions*, edited by Holmes, Liverpool UP, 2012, pp. 1-12.

Kac, Eduardo. "Biopoetry." *Kac Web*, www.ekac.org/biopoetry.html. Accessed 19 Sept. 2017.

————. "GFP Bunny." *Kac Web*, http://ekac.org/gfpbunny.html#gfpbunnyanchor. Accessed 6 March 2020.

————. *Hodibis Potax*. Édition Action Poétique, 2007.

————. *Holopoetry: Essays, Manifestoes, Critical and Theoretical Writings*. New Media Editions, 1995.

_____. "Media Poetry and Language Art. *Kac Web,* www.ekac.org/media.html. Accessed 19 Sept. 2017.

_____. "On the Notion of Art as a Visual Dialogue." *Art-Réseaux,* edited by Karen O'Rourke, Université de Paris I, 1992, pp. 20-23.

_____. *Recent Experiments in Holopoetry and Computer Holopoetry.* vol. 1600, SPIE, 1992. International Symposium on Display Holography.

_____. "Space Poetry." *Kac Web,* www.ekac.org/spacepoetry.html. Accessed 19 Sept. 2017.

Kalenberg, Angel. "Eduardo Kac: The Artist as Demiurge." *Art Nexus,* vol. 7, no. 69, June-Aug. 2008.

Langer, Susanne. *Feeling and Form.* Routledge and Keegan Paul, 1953.

Lennon, Brian. "Screening a Digital Visual Poetics." *Configurations,* vol. 8, no. 1, 2000, pp. 63-85.

Lenton, Timothy M. "Gaia and Natural Selection." *Nature,* vol. 394, no. 6692, 1998, pp. 439-447.

Levine, George. *Dying to Know: Scientific Epistemology and Narrative in Victorian England.* Chicago UP, 2002.

Leydesdorf, Loet. *A Sociological Theory of Communication: The Self-Organization of the Knowledge-Based Society.* Universal, 2001.

Livingston, Ira. *Between Science and Literature: An Introduction to Autopoetics.* U of Illinois P, 2006.

Lovelock, James. *The Ages of Gaia: A Biography of Our Living Earth.* Oxford UP, 1988.

Lynch, Lisa. "Trans-Genesis: An Interview with Eduardo Kac." *New Formations,* no. 49, 2003, pp. 75-90, www.ekac.org/newformations. html.

Maturana, Humberto. "Biology of Language: The Epistemology of Reality." *Psychology and Biology of Language and Thought: Essays in Honour of Eric Lenneberg,* edited by George A. Miller and Elizabeth Lenneberg, Academic Press, 1978, pp. 27-63.

Maturana, Humberto, and Francisco Varela. *Autopoiesis and Cognition.* D. Reidel, 1980.

Medeiros, Rafael Rubens de. *Ecopoesia dos novos meios: Eduardo Kac.* 2011. Universidade Estadual de Paraíba, MA thesis.

Middleton, Peter. "Cutting and Pasting: Language, Writing and Molecular Biology." *Science in Modern Poetry: New Directions,* edited by John Holmes, Liverpool UP, 2012, pp. 38-54.

Midgley, Mary. *Science and Poetry.* Routledge, 2001.

Noble, Denis. *The Music of Life: Biology beyond the Genome.* Oxford UP, 2006.

Pallardy, Richard. "Eduardo Kac: Brazilian American Artist." *Encyclopaedia Britannica,* 29 June 2019, www.britannica.com/biography/Eduardo-Kac

Price, Rachel. "Object, Non-Object, Transobject, Relational Object: From 'Poesia Concreta' to 'A Nova Objetividade.'" *Revista das Letras,* vol. 47, no. 1, Jan.-July 2007, pp. 31-50.

Rabinowitz, Lauren, and Abraham Geil. "Introduction." *Memory Bytes: History, Technology and Digital Culture,* edited by Rabinowitz and Geil, Duke UP, 2004, pp. 1-19.

Rose, Frank. "A Space Odyssey: Making Art Up There." *New York Times,* 23 Mar. 2017.

Rose, Steven. *Lifelines: Biology, Freedom, Determinism.* Penguin, 1997.

Rutherford, Adam. *Creation: The Origin of Life & the Future of Life.* Penguin, 2013.

Saunders, Lesley. "Do Poetry and Science have Interesting and Important Things in Common? Some Thoughts on 'Parsimony' and 'Provisionality.'" *Interdisciplinary Science Reviews,* vol. 39, no. 1, 2014, pp. 6-20.

Schleifer, Ronald. *Intangible Materialism: The Body, Scientific Knowledge and the Power of Language.* U of Minnesota P, 2009.

Stenlund, Sören. *Language and Philosophical Problems.* Routledge, 2003.

Vanderstraeten, Raf. "Rewriting Theory: From Autopoiesis to Commu-
    nication." *Systems Research and Behavioral Science,* vol. 29, no. 4,
    2012, pp. 377-86.

Varela, Francisco J. "Autonomy and Autopoiesis." *Self-Organizing Sys-
    tems: An Interdisciplinary Approach,* edited by Gerhard Roth and
    Helmut Schwegler, Campus Verlag, 1981, pp. 14-24.

All Systems Go on the Potter Wheel

RICHARD MURPHY

The astronauts on the Third Stone space craft
reach out to one another and barely touch.
With gratitude the evolved monkeys sigh
at any connection at fingertips.

Harmonizing behavior codes in room enough,
sol stewards watch for pitch, listen for bounce,
assisting fellow choir members who swing and twirl.

Interior states and organ representatives report
on mind and body functions: All A OK.
Roger everywhere, from toes to nose.

Oceans wave, sending friendly tidings
from frozen poles, from re-assured shores.
Mice and manes and fins feather in instinct.

Exhausted dinosaur ghosts that once roared by

from tailpipes rest in peace... again
finally with rust heaps in junkyards.

Having survived the planet pruning,
rocket scientists debated and won
against CEOs so space cadets could climb
into the nose cone cockpit
to smooth out the spin and travel
into the unknown gaps among stars.

Climate Climax

For a human being,
SUCK for venom and spit
to take the sting from a bee.
Introduce with all CAPS
but plan for anticlimax.
An S with fangs snakes
around town to deflate the alarm.
The ladder, engine and rescue squad
swarm about a cat in a tree.
"The situation could have been worse:"
An old sweet song
written by a philosopher.

However, should the sun
beat up skyscrappers
and gridded street fighters

or sea water top off already
stilted houses with people inside,
scientists jump up and down
on granted funds: no fun.
Money makers roll eyeballs
and jingle small change.
Solar System Pequod
doesn't turn on a dime
or for a silver dollar.

20/20 Vision

In the greenhouse, young corporate
candidates ripen into elite smug ignorance
on the lobby vine and on campaign trails.
Around the bend on a hill a white house
fertilizes and waters with history books.

Americans vote for nature
but live in disinfectant bubbles.

To pick an organic tomato not under glass,
radical taste buds need to bloom.
And only extreme weather, poisoned people,
debt-enslaved unemployed graduates
possibly blossom into placard-petaled streets:

Hong Kong throng, Moscow minion movement,
an Arab spring ring with crocus for peace
jarring the DC shock and awe windshields
and the NYC and LA la-la bell jars.

Between where the wild things scar
and hegemony doesn't care,
ideas enlighten to cultivate for a garden.

Western Hemisphere Adult Education

With an undiagnosed military industrial complex,
the useful idiot holes up in a basement bunker
eating canned psychology and a dried Eisenhower era
beef and sipping from a shiny TINA canteen.
The expert in a dunce cap waits to wise up
while the reinforced concrete media, Ayn Rand library,
and camouflage paint the united state.

Outside, Schadenfreude dams up allies, friends,
and the relative ease that once allowed Ahab to shoot
fish in a barrel while talking block party politics.

The boomerangs pivot over shrinking oceans,
over border walls, and over defense contracts.
In a too-late dream, manufactured cows graze
and derricks drill on the passing thought

"Avenue MLK/RFK" that never moved
off the childhood drawing table.

Hate grows on family trees in the many bombed out
distant lands mined beneath homes.
Once blindfolded Boston suburbanites lined up
against plastic straws where irony may soon surprise.

Wild

MATTHEW SHENODA

"Wild for life" as Lorde once said,
our futures broken
like the bones of lessons.

A prayer for the right kind of mending
a limp only noticeable in memory.

We are wild like this.

Wild in our ways of knowing.

Wild in the sandstorm of a forgotten sea.

Wild in the blooming on our kitchen tables.

Wild from the onset of winter.

Wild in our desire for flight.

Wild for life
until we tame and tame
dam the rivers of our freedom.

Wild for life
the way the wind hollows our lungs
gives us room for more.

Wild for life
even when dragged by the
trudging mud of memory.

Wild for life
the way our food once was,
sustenance from old ways.

Wild like the song
that makes us whole again.

Wild like this.

Wild like the spidered patterns of the coral reef.

Wild like the infinite crevices that shape our way.

Wild like the water we breathe.

Green Road

MARGARET RONDA

we passed it on the train going north
or on the highway
we rode it a long stretch
wild fennel in the air, egrets fishing
the silty shallows
felt along its edges
for the direct path we'd misplaced
a garden walk
we missed having missed the train
dashed out of breath
bluebells, primroses, neat borders
of the north country
or the end of the Green Line
with cattails and daisies sprouting up
just beyond the commuter parking lot
where the new encampments are

bright blue tents and cardboard
beds, a few shopping carts
sleepers, waking

it's some faraway place
that we drew nearer to, as if
we could relax and be
free
one state and then another
you have to drive through
past tidy suburbs where
no one sleeps well
foreclosed subdivisions at the
minimum wage you can't afford
to keep
alive in the rumble of sky
trucks passing on the right
bridges, corridors
stink of hot asphalt for miles

we found ourselves
near irrigation ditches and fields
of glyphosate
hoping to see things grow
we might have sung
the old song
how did it go

as the dew flies over the
green valley
no one knows the words
for a place with no weather
only heat and dust
daybreak and the road leads
to the industrial park
the fountain dried up

it's a poem about a green
path
strewn with goose feathers
where the thrush calls in the thicket
before the great war
we had to memorize the words
in grade school
an old man, a child, someone whose tracks
lead into the forest but never return
all things forget the forest

someone was holding hands
sharing a smoke in the alley behind
the burnt-out warehouse
we were on the surface
road at city's edge watching for
signs
for a wild grown-over

lane running behind the street
to the gated community
leading to the water tower covered in
graffiti, tiger lilies
spraypainted over names
where two men
nod off to sleep under tarps
amidst a sudden downpour

along the rural route through
the lofty hedgerows
we pursued with our eyes
in the yellowed photograph
hung in the lamplit musty village hall
where we arrived searching for
a mode of transport
to carry us bodily
backward or toward
the end of the
end of the century
you could hit the brakes and skid
all the way
crackling sounds in the corner
where the fire is being lit
by the old keeper

young green palace
green pleasant green
bowling green green ridge
shadow green village
green
run-down cul-de-sacs
close to the four-lane highway
the chemical plant
there's always a pipeline
leaking into the soil
some kind of dust settling
on the dark leaves
wherever we woke
the road carried us

Cyanotype in a New York Public Library

VICKIE VÉRTIZ

The amount of work that Anna Atkins—with the help of
her servants–
had to do over the years to produce
British Algae should not be underestimated

How do you press the earth into a book?
Glass planes turn salty sea plants into hair fractures
A carrot top with a sun in its heart
Chestnuts with looping tails
A feathered wig for a Who

Here I am, dredging up olive trees
underneath a sky painted on the ceiling
A window–laden room

The most beautiful thing, the most beautiful sound
is a library full of silent readers. Wooden chairs scrape
the floor. The occasional cough. Pages being turned, my
hands
covered in topaz baubles

The leaves might stick if they are too mucousy
and then there is no image to reveal. Miss Anna
Which are your ideas and which were someone else's?
What is your work and what is someone's life?
What is your hobby and what is my survival?

Servants are scientists and actors
Here is me pretending to be interested in sea trash
Here is me instructing my maiden on her indigo backwash

In the Amazon there is still war
In the name of cleansing, we burn palo santo
which are our ancestors. We rob them of their shade for
good

Santa Ynez mountains are ablaze with temperature and
powerlines
And yet, the stolen water and grass baskets
found the last family who knows their songs

I turned another page of azure and old debuts
This one is missing its jellyfish tentacles
That one is a wiry white beard

Comadre, I don't know what to tell you
We are with you, in different cages, tipping into a new
page
Here is a spade, a roof. Flat grasses and two almonds

If Guinea fowl feathers could speak for themselves they'd
want
a future of playing in violet sheets and daylight
Polka Dot feathers and tiaras made of amethyst

Hydrolytic tendency means I have a mildly acidic condition
So I guess I am a sour ass motherfucker

A servant in service to my writing, I'm an electron
of this marble, a sedimental lion if there ever was one
Eyeshadow on this place from now on

The Results Are Spreading

MAURICIO KILWEIN GUEVARA

Dead man's bells, Digitalis, Fairy cap, Fairy finger, Foxglove, Lady's thimble, Lion's mouth, Purple foxglove,
Scotch mercury, Throatwort, Witch's bells. Which is
consistent with the finding that the foxes have fallen
there. Unlike the kumiho who can live a thousand years.
Or the nine-tailed fox, changeling of a thousand tales. The
termites have gone on another book tour, glowing as they
go. The fish, however, are agnostic and resist reading as
a rule. Nevertheless, the massive culling is already under
way. The CAFOs are heating up, shitty business but you
can make a killing. The results are now ready. As with
last year's report, key findings are still being redacted
to ensure national security targets are struck. The data
regarding the cranial thinning of human infants have
been excised from all Excel addenda to this report. The
fox grape and wisteria are perpetually worried. Even the

morning glories at happy hour set their teeth on edge.
The gypsy onions have wandered off again. The silver
lining: the egrets and the herons have begun to organize.
To brush, the long slender feathers cascading down the
other's back during the breeding season. Let us praise the
action of the egrets and the bitterns – who are spreading milt and roe as they wade through the rising waters.
Meanwhile, Amitav Ghosh remains under house arrest for
regretting that *The Last Days of Snow* would be returned to
his agent with the following note: *The revolution will not be
novelized. Dig where you stand. The difference is spreading.*

Peace Over Everest

WANG PING

above 20,000 feet, all extras must go:
camera, clothes, jewelry, food, urine, feces
blood stops flowing to fingers, toes, limbs
to liver, stomach, pancreas, intestines
only the lungs stay to bring air
to the heart that pumps blood
to the brain that keeps
the will alive

this is letting go
our daily mantra for deeds
yet we cling to worry, anger, hate
toxins drowning us like flashfloods
we attach to beauty, memory, love
till they become fossils of fear
plaguing our arteries

cancer arrives when cells refuse to die
madness bursts if we cling to our desire
disease comes if we don't let go food or love
death seizes us all as we chase after immortality

yet no star tells the same story of the universe
no mountain stands still on the earth
no snowflake forms exact patterns
no finger coils into same prints
no tree grows same rings of life

so let go
the fire for this world
and sow it as seeds of forests
melt the knives of fame and power
into ploughs to cut through fields of ego
let go our clutch to anger, hate and fear
turn them to fertilize flowers of spirit
let go all extras...only the lungs stay
to bring air to the heart to pump
blood to the mind to turn
pain into joy and hate
to love and death
to life and life
as peace

Fireblossoms

AKI GIBBONS

Vining through our days
and illuminating dreams,
a sinuous undercurrent—

at times emerging as
a deity
a story
a fear
a longing

a whisper of a spark
a foreshadowing
spreading like veins through flesh
humming for the next season

inside us, a flowering of wordless knowing
yet we're bewildered witnesses of chaos

of a withering within
and every day the garden looks
stronger lusher deeper

clearer, seeing glimmers of what we
thought were ours alone
now outside,
yearnings taking form,
words taking leaf,
wildness with a face
turning slow-blinking eyes to us

(this world is collapsing and we continue working—
from our rooftop office in the middle of the city
we watch a glass tower being rapidly devoured
by violently beautiful vines,
blossoms erupting forth like
vibrant fireworks)

we take photos as the tower slowly goes
down, down, down

and now the dream and dreamer become one.

Ode to surfaces, air

ORCHID TIERNEY

What kind of monster is I park discreetly outside the former Model Landfill, on the driveway close to the two-lane road, far enough from the entrance to appear that I am lost. The landfill opened in 1967, and it was *an anomaly, he said. This is not something that happens on any regular basis* one of the first sanitary landfills to operate in Columbus, Ohio.

Located off Jackson Pike and west of the Scioto River, which runs through the city, the Model Landfill was built during a period of intensive professionalisation in the U.S. waste management industry. Seven years earlier, Columbus sanitation workers became uniformed employees of the newly organised Division of Sanitation for the first time in the city's history. Military men. Their outfits were beautiful and cathedral, cathedral because they were beautiful men.

The purdu leakages are malicious and tasteless. The Model Landfill employed loving men. It covered one hundred acres of industrialised land and accepted commercial and residential wastes. It operated continuously until its formal closure in

1985, and the Solid Waste Authority of Central Ohio (SWACO) took over management of the site. The form of the land scattered in deeds and documents: SWACO leased the site to Phoenix Golf Links Ltd.–Petro Environmental to construct a golf course over the capped landfill in 1999.

This alliance between pleasure and bioremediation is a common green narrative. It replicated similar urban engineering projects at FreshKills on Staten Island, Spectacle Island in the Boston Harbor, and Tifft Nature Preserve in Buffalo.[1]

The Phoenix Links Golf Course opened in 2000, *and we do our best to prevent this from ever happening,* at first blush, it exemplified successful brownfield remediation.

It was sweet and fitting. As Christopher Todd Anderson notes, "Finding meaning in garbage requires a departure from the usual sites of human activity and a willingness to enter the ambiguous space of the dumping ground or the littered field."[2]

But while waste reimagines the scales of human leisure in the toxic traces of consumer culture, the ambiguous spaces of the Model Landfill are molecular and effervescent.

I stay for only a near half a minute. My photos of the site's entrance grasp the frame but ignore its depth *none of us pay attention to that which we can't see or measure.*

A creek spills brown water from a pipe at the base of the landfill: a heap of tide bottles and language; plastic shrimp; a tang of sewage, and hardened veils of glass. *These ghastly airs partially resist their enclosure.*

the elastic surfaces of molecules, light

Methane is an incoherent lyrical subject. Odourless, invisible, unwinding wood wine. Methane migrates, withholding its burial. *We've got it under control. We're managing it.*

In 1774, Benjamin Franklin wrote a letter to Joseph Priestley, the discoverer of oxygen, about a mysterious "inflammable gas" found in woodlands and swamps. In that letter, Franklin described secondhand observations gleaned from amateur experiments he had overheard while passing through *39.864°N, 80.861°W* New Jersey. "I heard it several times mentioned," he wrote, "that by applying a lighted candle near the surface of some of their rivers, a sudden flame would catch and spread on the water, continuing to burn for near half a minute."[3] *A spark bears witness to frightful airs.* A spark is not a surface. *It was red for days, relentless.* The integument had exposed itself. While Franklin noted that others had replicated these experiments, he had been far less successful. It is reasonable to assume that this *fire* gas *lasted for three days* was none other than methane, bubbling from New Jersey's hebetic bogs. Priestley later published the letter in "Experiments and Observations on Different Kinds of Air." Franklin's account of these magical flames, ignited from the muddy surfaces of marsh water, also inspired the Italian physicist Alessandro Volta to identify the ghostly substance. *These ghostly airs partial and resistant.*

Volta managed to isolate methane from the marsh gas he had meticulously collected from Lago Maggiore, Italy's second largest lake, in 1778. Unlike an ode, there is no straightforward story here. An ode is a surface, sweet and fitting. Naming conventions for this fiery air wouldn't become standardised until the nineteenth century, when August Wilhelm von Hoffman, a German scientist, derived the name released more methane ($CH4$) from 'methanol'[4] *than heavily industrialized European countries like Norway and France.* A name is a kind of sticky surface.

Methane is grey scholarship. Classical odes are monstrous and rigid. Wetlands, marshes, and oceans are natural sinks of methane.

Hydraulic fracking of the Marcellus Shale, which covers an area from New York to the Appalachian Basin, releases fugitive methane emissions into the atmosphere. One study, using plain language, estimated a "mean emission rate of ~21 kg $CH4/s$... from a 4,200-km^2 study area [in the southwest region of the Marcellus Shale]. A significant portion (~70%) of the emitted $CH4$ was found to originate likely from coalbeds."[5]

Data is form and knowledge.

I don't know what this means. The release of methane *Gas! Gas! Quick, boys! – An ecstasy of fumbling*[6] from organic sources like the Arctic permafrost is a terrifying contributor to global warming. Methane is greenhouse knowledge, and its *the blowout expelled sixty thousand tons of methane into the atmosphere* emissions are also difficult to track, since the gas has a tender lifespan of 9.1 years in the atmosphere (compared with one hundred years for carbon dioxide).

But *the global surface temperature is my skin* methane brought me to the Model Landfill: this gas is a natural by-product of decomposing garbage. Landfills are sentient and restless, geological forces as dynamic as volcanoes. As Kathleen M. Millar writes, "Garbage is gaseous, belching methane and carbon dioxide that must be trapped and released lest the methane spontaneously erupt into fires."[7]

According to the Environmental Protection Agency, landfills accounted for 14.1 percent of methane emissions in the United States in 2017.[8] The hourly rate of emissions exceeded the Aliso Canyon event. The collection of methane or landfill gas (LFG) is therefore vital to maintain healthy air quality, especially for residential and industrial areas that surround these modern wastelands. Vertical and horizontal pipes, buried underneath landfill cells, collect LFG and transport it to processing facilities where *a mere one mile away, residents were swiftly evacuated* it is either burned off or converted into electricity, heat, or fuel for residential and commercial energy use. *An archive discloses.* Methane defies the inefficiency of commodity culture; it is the retaliation of the contagious rejected and a rejection of climate repair.

I can't portrait this: this conversion process of methane underscores the transitory nature of landfills. Gaseous wastes are mobile, leaky, and porous commodities, traveling far from their initial sites of deposit, returning back to cities, homes, and commercial spaces as ghastly vapour.

Indeed, methane foregrounds how wastes are not "matter out of place," as Mary Douglas imagines, but are very much 'in place' in urban society.[9] *The archive encloses.*

Landfills may be sites of "managed decay"[10] that naturally conserve wastes as corpses of menacing record, but methane resists these archival tendencies. Look at this surface. It is crooked and solid. It is impossible to write a poem about pastoral repair. Methane cannot repair poetry.

the film partially develops, leaving the pastoral open to fancy

I inhale the data of air. Exhale. After methane emissions were discovered at the Model Landfill, SWACO appealed to the courts to break its lease with Phoenix Golf Links, claiming that the golfing company had insufficiently managed the landfill gas system. The golf course closed permanently in 2015, and SWACO shelved plans to reopen it at a later date. I am here in 2019.

This is not a praise poem. *Language cannot do everything*[11] – This is gaseous modernity. When a gas attack violently startles Wilfred Owen in his famous World War I poem "Dulce et Decorum Est," all the men in his company, bar one, swiftly don their masks. The slowest soldier suffers an agonising death: "Dim through the misty panes and thick green light," writes Owen, "As under a green sea, I saw him drowning."[12] *Some people weren't able to return to their homes for weeks.* The green mustard gas gasps his failure, suffocates the trench pastoral, and his terrifying death asphyxiates the poet with nightmares.

This toxicity seethes verdurously in the poem. It is gaseous. It is ugly vapour. *A gas exchange in mammals.* The gas attack is not an ode to the necropastoral – what Joyelle McSweeney describes as "the manifestation of the infectiousness, anxiety, and contagion occultly present in the hygienic borders of the classical pastoral,"[13] but I am attracted to this idea. I look for the idea of surfaces, air at the Model Landfill.

You tend not to address what you don't know. Owen's gaseous poetics violate the shepherd's imaginary of fecund meadows and tender. This was the part of the ode he painfully disclosed. *The event was an anomaly until it happens again.*

On the driveway, *I deeply regretted the incident* take photographs of trees and signs, tide bottles and language for near half a minute. One photograph partially develops, leaves a crooked line between appearances and fancies.

A line is not a surface. The allure of the image is what it refuses to name: the slanting outside of the hasty frame, the ugly substance of stuff, the ghastly airs partial and present. That's the model story of the landfill: a golf course is not an archive. Print the light, surface, air. Pray the depth sticks.

Endnotes

[1] "ode to surfaces, air" is an anti-ode that interrogates the invisible voices of disaster through the optic of methane gas. The Belmont County, Ohio, methane explosion of February 2018 has largely gone unnoticed in the media, although the full scale of the leak was finally released to the public in December 2019. By pairing a secondary lyrical voice with my formal investigations into a local landfill–turned–golf course, which had itself been closed down because of fugitive methane emissions, the anti-ode yields to the limits of witnessing – and knowing with certitude – the magnitude of air pollution.

See Scappettone, Rueb, and Skinner.

[2] Anderson, p. 50.

[3] Priestley, pp. 321–23.

[4] Hofmann, pp. 55–62.

[5] Ren et al., pp. 1862–78.

[6] Owen, p. 55.

[7] Millar, p. 31.

[8] See EPA.

[9] Douglas, pp. 36.

[10] See Gabrys, p. 107.

[11] Rich, p. 19.

[12] Owen, p. 55.

[13] McSweeney, p. 3.

Works Cited

Anderson, Christopher Todd. "Sacred Waste: Ecology, Spirit, and the American Garbage Poem." *Interdisciplinary Studies in Literature and Environment*, vol. 17, no. 1, 2010, pp. 35-60.

Burger, Beth. "Ohio Gas Well Blowout Released More Methane Than Some Countries Do in a Year, Researchers Find." *Columbus Dispatch*, 27 Dec. 2019. www.dispatch.com/news/20191223/ohio-gas-well-blowout-released-more-methane-than-some-countries-do-in-year-researchers-find.

Douglas, Mary. *Purity and Danger: An Analysis of the Concepts of Pollution and Taboo.* Routledge, 1966.

EPA (U.S. Environmental Protection Agency). "Basic Information about Landfill Gas." www.epa.gov/lmop/basic-information-about-landfill-gas. Accessed 14 Dec. 2019.

Gabrys, Jennifer. *Digital Rubbish: A Natural History of Electronics.* U of Michigan P, 2011.

Hofmann, A. W. "On the Action of Trichloride of Phosphorus on the Salts of the Aromatic Monoamines." *Proceedings of the Royal Society of London*, vol. 15, 1866, pp. 55–62.

Jarman, Josh. "Trash Authority Gives Up on Landfill Golf Course." *Columbus Dispatch*, 13 Mar. 2015. www.dispatch.com/article/20150313/news/303139760.

McSweeney, Joyelle. *The Necropastoral: Poetry, Media, Occults.* U of Michigan P, 2015.

Millar, Kathleen M. *Reclaiming the Discarded: Life and Labor on Rio's Garbage Dump.* Duke UP, 2018.

Owen, Wilfred. *The Collected Poems of Wilfred Owen.* New Directions, 1965.

Pandey, Sudhanshu, et al. "Satellite Observations Reveal Extreme Methane Leakage from a Natural Gas Well Blowout." *PNAS,* vol. 116, no. 52, 2019, pp. 26376–81, doi:10.1073/pnas.1908712116.

Pfleger, Paige. "A Fracking Explosion in Ohio Created One of Worst Methane Leaks in History." *WOSU Radio,* 19 Dec. 2019, https://radio.wosu.org/post/fracking-explosion-ohio-created-one-worst-methane-leaks-history#stream/0.

Pinckard, Cliff. "Satellite Data Shows Ohio Methane Leak One of Largest Ever Recorded, Reports Say." *Cleveland.com,* 17 Dec. 2019. www.cleveland.com/nation/2019/12/satellite-data-shows-ohio-methane-leak-one-of-largest-ever-recorded-reports-say.html.

Priestley, Joseph. *Experiments and Observations on Different Kinds of Air.* Vol 1. London, 1774.

Ren, X., et al. "Methane Emissions from the Marcellus Shale in Southwestern Pennsylvania and Northern West Virginia Based on Airborne Measurements." *Journal of Geophysical Research: Atmospheres,* vol. 124, 2019, pp. 1862–78. doi:10.1029/2018JD029690.

Rich, Adrienne. *The Dream of Common Language.* W. W. Norton, 2013.

Rueb, Teri. "Core Sample, Art on the Harbor Islands." June 23–Oct. 8, 2007. Boston: Spectacle Island & ICA Founders Gallery. www.terirueb.net/core_sample/index.html.

Scappettone, Jennifer. *The Republic of Exit 43: Outtakes & Scores from an Archaeology and Pop-Up Opera of the Corporate Dump.* Atelos, 2016.

Skinner, Jonathan. *Birds of Tiff.* BlazeVOX, 2011.

Tabuchi, Hiroko. "A Methane Leak, Seen from Space, Proves to Be Far Larger Than Thought." *New York Times,* 16 Dec. 2019. www.nytimes.com/2019/12/16/climate/methane-leak-satellite.html.

RUN

CAMILLA NELSON

For the last ten years my practice has investigated the particularities of how words develop from an embodied and enmeshed human experience of being with and through the world. I have worked with dancers and sound artists to develop my understanding of how the moving sounding body responds to and generates its environment in a more intensely physical and sonic capacity, and as part of this investigation, I took a workshop in spring 2018 with the physical theatre group OBRA at its headquarters in France, with the British poet and fell runner Lucy Burnett.

The sequence that follows is the foundation from which my most recent project, RUN, develops. RUN is an exercise in exploring how the moving body thinks and feels its way through an animated being in/with/through landscape. It was inspired by a prompt from Lucy Burnett, during the OBRA workshop, to write a response to running. We were driven to the top of a short hill and invited to allow ourselves to fall as far and as fast as we could down the hill—putting your legs out to catch you only when you really needed to. The first time we ran down the hill with our notebooks and pens in hand, and on arriving at the bottom we immediately spilled whatever words came out onto the paper, entirely uncensored and disorganised, just documenting what came. We were then instructed to leave our notebooks and pens at the foot of the hill and walk back up to run (or fall) down as before and let the words pour out of us upon reaching our books at the bottom of the hill all over again. We repeated this process five further times.

This intense burst of physical activity gave rise to exactly the type of language that I am always striving to achieve, a stream of words that are of the landscape and my being with and through each other at the same time, without any sense of division between the two. This experiment generated an expression of enmeshment and co-emergence that I had been trying to achieve in many different ways, and following my return home to England, I was inspired to take it still further over a greater distance and a longer time period.

For six months, I applied this method of running and writing as a way of documenting the emergence of self through, by, and with the landscape where I live. From May to October 2019, I ran from my house at the foothills of the Mere Downs up onto the top of the downs and around the craving hill that surrounds the town, dropping back down, into the town and back to my house. The rule was slightly adjusted from Burnett's first prompt. This time the rule was to run until you couldn't run any more, then write. When I stopped running, I wrote until I could run again. When I wasn't writing I was running, and when I wasn't running I was writing. The writing was whatever came into my head, as it comes—as uncensored and immediate as possible. The focus is not to produce a beautiful piece of writing but to track the process of how words form while running/writing a particular landscape.

This same process was repeated over the same route every month for half a year. The aim being to track not only the emergence of my inner landscape and my embodied relationship with this landscape but to document the combination and evolution of all these things and the landscape's rhythms beyond me, as it shifted from spring through summer into autumn, moving towards winter.

My focus was to get away from the impulse to describe as if I were an "outside" observer. As we are all plainly aware, there is no outside this multiply inflected and intricately combined series of ecosystems we emerge with, by, from which to write. RUN is an attempt to give an account of this emergent process of being with/in language/landscape, and it stems from these first six hill runs inspired by Lucy Burnett.

I invite you to attempt your own running/writing experiments to access your own particular sense of being with/in the world through language/landscape.

|

breathe cavities form lungs

push up grass   from breath

shit song   from every orifice

||

arm-wheeling umbels

coniferous restraint

wolf gallop & loll

teeth glisten

lip quiver

heave

heave

heave

locust song

|||

snap dragon   sweet pea

twist & roar   turnpike emphasis

cattle grid   rattle   gear   brake

off peak   neck   neck

speed is   a grizzly bear

                         making complaint

hungry   for revenge

fly catcher   neck wrenched

opposites   make friends

IV

limb-like structure of grasses

bite round your ankles   ticking

with the fizz of falling seed

machine bleat surge   mechanical

outrun by other feet   trimmed

to cut and bleed   when senses arrives

cease   block the thread

                        cut the tongue

                        crop the breath

V

and plop into a sediment of green-tongued roughage

thread the song of broken breath   entwine

your seed head   bend it   towards mine

pollen heavy   open your eyes

to the lash-like   interior

of networked veins   train   breath to become

                                                          lighter

draw
                        draw

                                          till the flame no
                                                    longer
                                                       resists

VI

sweet angled speed   adrenal

pant your way out of   underground

obvious   tubular windpipe

water-horse irrigation

birds chirrup you

out of your   blind interior

underwater   drowning for a sense of

earth beneath   your

what about falling?   nobody mentioned that

or   finding your own way home

at night   with all the lights blown out

nobody mentioned that

what it comes down to   is   this

how do the stones sing out?   in clover

& tinselling firework fronds

Grosse Fatigue: Eco-Weariness and the Feminine Hand

EVELYN REILLY

Flowers

Having a taste for crazily dense and ambitious cross-genre work, I was intrigued by a show I walked into almost by accident a few years ago of work by the French artist Camille Henrot. Her video piece *Grosse Fatigue* had made a splash at the most recent Venice Biennale, but the first room at the New Museum in New York City was filled entirely with flowers.[1] The installation consisted of a series of floral arrangements, each associated with a book she had left behind when moving from Paris to New York. A vase of white orchids with deep pink throats was labeled *The Black Book* by Lawrence Durrell, Gertrude Stein's *Composition as Explanation* was present in the form of a cluster of pleasantly garish red and yellow mums, a stately palm leaf represented Aimé Césaire's *Discourse on Colonialism,* and so on. My personal favorite was John Cage's *Diary:*

How to Improve the World (You Will Only Make Matters Worse) – a head of radicchio sitting on a plain round base. The piece as a whole was titled *Is it possible to be a revolutionary and like flowers?* While various reviews of this piece asserted connections to the material, economic, and political histories of individual flowers and reminded the viewer of the floral monikers of "revolutions" of recent decades – the Rose Revolution in Georgia, Jasmine in Tunisia, and others – I was mainly struck by the fact that a woman artist had filled a gallery space with what were, in spite of the witty interventions, basically flowers in nice arrangements.

On one hand, I was interested in the overtly "feminine"[2] aspects of this project and the artist's willingness to court the usual ambivalences for what Susan Howe, in her book *My Emily Dickinson,* named "embroidery" – as in "the first labor called for was to sweep away the pernicious idea of poetry as embroidery for women" (17). By drawing on decorative traditions based in a domestic domain neither overtly subverted by irony (as in work of second-wave feminist artists such as Rossler or Schneemann) nor celebrated as the stage set of erotic action (long an accepted subject of "domestic" art by men), the artist risked engaging some of the ambiguous baggage of feminist art. So it was interesting to me that these pieces were by a fairly young and obviously ambitious artist. They were, unabashedly, *delicate.*

The installation also brought to mind the subject of "reading while female." Asserting a connection between flower arrangements and books, predominantly by men and pretty typical of the required reading of the art-academia nexus, Henrot touched on a subject I'd long pondered – the female mental inhabitation of a male subject position as a core experience of formative intellectual life. These thoughts had been rekindled at an event at St. Marks Poetry Project that called for a discussion of "the white room" of the poetry reading.[3] During this panel about the racial composition of poetry audiences, there was a moment when one of the participants mentioned having frequently inhabited the white subject position, even the white racist subject position, as a "normal" part of his education. While all this comes with the territory of an intellectual landscape littered with historical diminishments and absences, which the reader endures, even adapts to, in the name of larger aspirations and desires, the long-term impact of such psychic contortions can't really be other than a terrible *weariness* punctuated by more or less suppressed episodes of rage.

None of this was overtly addressed in these arrangements of "flowers as books," which were, from another point of view, simply assemblages of uprooted or sampled nature.

Familiar hybridized blooms, cultivated in an age of imperial obsession with the "exotic," were mixed with the "native" or "wild" – a posy of terms ripe for horticultural decolonialization. Neither Mallarméan "bouquets beyond bouquets" nor maudlin Baudelairean fleurs, these three-dimensional riffs on the "still life" tradition were rather very postmodern juxtapositions of natural materials and cultural references.

Not the artist's greatest work in my view, but interesting, I thought, in the questions it raised about artistic strategy, about the deliberate use of traditionally gender-associated aesthetic terrain (which I've sometimes thought of as the "redemption

of the feminine"), about mediated nature, and about the revolutionary gestures of art in our current moment. These are
issues that emerge as well in recent poetry by women, such as
Juliana Spahr's chronicle of her affair with "Non-Revolution" in
That Winter the Wolf Came, Brenda Hillman's *Extra Hidden Life,
among the Days,* Mei-mei Berssenbrugge's *Hello, the Roses,* and
Harryette Mullen's *Urban Tumbleweed: Notes from a Tanka Diary,*
to name a few examples from the United States.

Frogs and Phones

In the next room, Henrot's *Grosse Fatigue* was installed—a
thirteen-minute video piece whose title could be translated
as "Great Tiredness" or "Grand Exhaustion," but which also
chimes in my ear with the "Grand Fugues" of European musical tradition.

Using the conceit of the computer desktop as compositional frame and incorporating film sequences shot behind
the scenes at the Smithsonian Museum in Washington, D.C.,
the piece is indeed a densely interwoven fugal composition
of images drawn from art, anthropology, and natural history.
Drawers upon drawers of scientific specimens appear, as well
as airline magazine covers from the 1960s, moonshot posters,
various Wikipedia pages, a Darwin bobblehead doll, and a
woman, the artist perhaps, with her own hand in her crotch.
All of these are "displayed" via opening and closing browser
windows that manage to designate the contents both as "art"
and the "not-art" of the digital cornucopia.

A spoken word poem created by the artist and her poet collaborator Jacob Bromberg accompanies these visuals, drawing language from what could be characterized as religious, pagan, and scientific "origin of species stories," with aesthetic echoes of The Last Poets of the 1970s, the poetry of Inger Christensen, and New York School list poems. One long sequence shows image after image of representative plants and animals, accompanied by a recital of the scientific names of the groups to which they belong.

While this piece is quite different from the questionably revolutionary flowers, there are also continuities—the interest in institutionalized knowledge systems (the personal library, the museum, the encyclopedia, and their digital counterparts), the not shrinking from some conventional notions of beauty, and also the presence of the feminine hand in both literal and aesthetic senses. The nails are *polished;* the gestures *graceful.*

In an interview, Henrot said "the film is an experience of density itself" (qtd. in Simonini), which I think relates to an aesthetic of immersion and permeability and even saturation – one of Henrot's images being the squeezed-out sponge – highly relevant to our current circumstances, among them the globalized world of cultural assets and the digital space through which we try to navigate them.

In even broader terms, the work evokes an experience of "superpermeability," or maybe it's better described as "juxtapositional excess." It enacts a kind of extreme interrelatedness relevant to contemporary poetics, including so-called ecopoetics, which I prefer to think of as simply a poetics awake (woke?) to the necessary reconceptualization of the meaning of human life in the context of our species' role as agent of mass extinction and climate destruction. I remain convinced that if in poetry or in any other artistic sphere, we consider the ecological as merely a "genre," one on a list of more or less interesting or hip stances, instead of as an exploration of new ways of "thinking the world" (or, to put it in political terms, if climate change is seen as one issue on a list of issues one cares more or less about), then we are indeed doomed.

It's not that I think poetry can be particularly effective in this; I don't. And it's not that work produced by poets calling themselves eco-poets is particularly good. But I continue to find it strange, if not borderline psychotic (and I mean this in a collective sense), to produce art that remains unmarked by a cataclysm of a scale that will likely make the ethical and intellectual reorientations caused by, say, the First World War seem small by comparison. So I am always looking for aesthetic responses that aren't just another gesture of ironic despair or deer-in-the-headlights nihilism.

Which brings me back to *Grosse Fatigue* and the crazily exhilarating "tiredness" that accumulates as the piece runs through its catalog of mostly exhausted cultural resources, including so many of the framing devices through which we have perceived our place in nature since the beginning of the Industrial Revolution. The Romantic landscape, the museum with its history as storehouse of colonial loot, the glossy natural history photograph, the poem, the YouTube video, all, in the end, amount to a "grosse amalgamation" that reflects both the euphoria of information age plenitude and its loneliness and antisocial sociality.

I also admire Henrot's success at incorporating, without actually containing, so much worldly "stuff" into her art and connect that to struggles I've had with the constraints of the lyric, the poem, even the book of poetry in terms of its ability to engage with this overflow, this overwhelming coincident connectedness and disconnectedness that are part of our current experience of being.

Two examples:

An emblematic image of the piece is of a small frog sitting on an iPhone and exquisitely color-matching a shade I think of as "celadon" (one the most highly cherished ceramic glazes of Chinese imperial culture). This is a nature about as mediated as it gets, posing, so to speak, on the ultimate industrially produced object of desire, polished to its own high degree of finish in one of the factory cities that both fulfill our material lusts and endure the environmental consequences.

A counter example is a faded poster that the artist probably photographed in a hallway at the Smithsonian. Announcing an exhibition, long dismantled, called "Art in the Service of Ichthyology," it serves as an apt statement of the failure of her own and all such art projects as they grapple with the state of the living world—that failure being, in the end, at the heart of the *great weariness* of the piece.

The Poem

A first draft of this essay was written a few years back in response to a call for papers for a conference in Buffalo, New York, which asked, "What is a poetics of the next 25 years?" My answer to this was that mostly I had no idea, but that some of what I appreciated in *Grosse Fatigue* might be among its qualities. In addition, I thought that whatever this poetics might be, it probably must include what I think of as a staring-into-the-abyss unflinchingness, a refusal to look away from the most difficult aspects of our present circumstances.

Thirty years ago, Susan Howe published *My Emily Dickinson,* in which she described a "nature" that was formative of the Puritan American psyche: "Emptiness without design or plan, neighborless in winter blank, or blaze of summer. This is waste wilderness. Nature no soothing mother, Nature is annihilation brooding over."

These early shock-and-awe encounters between settler-colonialists and native terrain would continue to have their resonances in European American fantasies of nature as "untouched" wilderness and blank slate for self-invention. For African Americans, the response was quite different to a nature that was a place of precarious refuge and potential freedom, but which, over the course of history, would become so infused with the memory of terror that no tree could ever be just a tree again—a seemingly permanent haunting, which is so evident in *Black Nature: Four Centuries of African American Nature Poetry,* edited by Camille Dungy.

For a late Romantic such as Dickinson, nature was a soul-shattering blank left behind by a departed God – an emptiness, into which she became an adept practitioner of the unwavering stare.

Now, 150 plus years after she wrote, "We came to Flesh – upon – ... Miles on Miles of Nought –" ("I tie my Hat – I crease my Shawl"), the envelope has been turned inside out. The figure of emptiness turns out, as it always does, to have been a cover for denial of dependence upon undeniable presence – more applicable to the blank of our social will than to the nature of which we are a part. And the "withdrawal of God" elegies that gave rise to so much nature poetry turn out to be pretty much beside the point at a time when nature even at its most red in tooth and claw seems almost hopelessly vulnerable in the face of the industrial wreckage of the planet.

So what might be a poetics of the next twenty-five years?

Maybe it's in "Those Evenings of the Brain—" when "... something in the sight/Adjusts itself to Midnight–," as Dickinson would have it ("We grow accustomed to the Dark–").

Or in Inger Christensen's prescient

> ...the specific
>
> contours of the moment show
> the future's embryo
>
> as a contagious stand of fossils
> (*Alphabet* 73)

Or Berssenbrugge's
> Person and violet with so little in common my voice re-
> veals as a resonance of unmanifest identity.

("Glitter" 47)

Or Hillman's

>   . . . Groups & nations
> howl unseen . . . The mind
> god-labors
> pumping itself green.

("The Forests of Grief & Color" 13)

Or Mullen's

> discarded plastic bag we see in every city
> blown down the street with vagrant wind.
> (*Urban Tumbleweed* 32)

It's probably also in Spahr's

All day long, like a lion I lie where I will with not really me
and I bestow upon not really me
refractive index testing oils and wood preservatives.
("Tradition" 56)

These lines, referring to her fear-laden care of her young
son, hark back to the toxic parables and environmental poetics of Rachel Carson, who, famously, finished writing Silent

Spring while ill with terminal cancer and then continued, in spite of her exhaustion, to defend it from attack.

May we continue in these weary footsteps.

Endnotes

[1] This essay is a revised and expanded version of a talk given at the conference Poetics: (The Next) 25 Years at SUNY Buffalo, April 2016. Camille Henrot. *The Restless Earth*, the New Museum, New York, May 7–June 29, 2014.

[2] I am using "feminine" here in the sense of an aspect of the performance of gender available to artists of any gender identification and in keeping with Joan Retallack's definition of "the experimental feminine" as a practice, aesthetics and poethics available to all poets. See Retallack, "Alterity" and "Experimental."

[3] "White Room," a discussion at the Poetry Project at St. Mark's Church, New York, January 2016, in response to Spahr and Young.

Works Cited

Berssenbrugge, Mei-mei. "Glitter." *Hello, the Roses.* New Directions, 2013, p. 47-50.

Christensen, Inger. *Alphabet.* New Directions, 2001.

Dickinson, Emily. "I tie my Hat—I crease my Shawl." *Emily Dickinson's Poems as She Preserved Them,* edited by Cristanne Miller, Belknap, 2016, p. 256.

————. "We grow accustomed to the Dark." *Emily Dickinson's Poems as She Preserved Them,* edited by Cristanne Miller, Belknap, 2016, p. 171

Dungy, Camille T. *Black Nature: Four Centuries of African American Nature Poetry.* U of Georgia P, 2009.

Hillman, Brenda. "The Forests of Grief & Color." *Extra Hidden Life, among the Days.* Wesleyan UP, 2018, p. 13.

Howe, Susan. *My Emily Dickinson.* New Directions, 1985.

Mullen, Harryette. *Urban Tumbleweed: Notes from a Tanka Diary.* Graywolf, 2013.

Retallack, Joan. "Alterity, Misogyny & the Agonistic Feminine." *Jacket2,* 6 Feb. 2018.

————. "The Experimental Feminine." *The Poethical Wager,* U of California P, 2004, pp. 90-101.

Simonini, Ross. "To Have and to Know: Camille Henrot's Elephant Child." *Art in America,* 2 June 2016.

Spahr, Juliana. "Tradition." *That Winter the Wolf Came.* Commune Editions, 2015, pp. 53-58.

Spahr, Juliana, and Stephanie Young. "The Program Era and the Mainly White Room." *Los Angeles Review of Books,* 20 Sept. 2015.

(that reflect heat back into space)

DANA ZOUTMAN

a clear pattern of underestimation
a realistic win-win
an emergent property
an undeniable reality.
Appalachian Mountains
at an abandoned factory in the town
blamed the incident on "very unusual temperatures"
change turned into a crisis
climate change is remapping the world
considered a lesser crime
consistent for decades
delaying a reasonable and orderly transition
deliberate deception
Don't we need to stabilize
dreadfully expensive climate inaction
fire weather conditions
focus on ourselves. Meditate and shop at farmers' markets
human-induced

I am in no position to judge
it manufactures evidence
it often fails the victims
justification for inaction
Knowing this, what do we do?
Living with this kind of cognitive dissonance is simply part
of being alive
need to be broken down
no longer resisted
northward into the southern
Now imagine those emergencies
offer emotional support
publicly air
rapid swings to the fire weather
recall flashes of the encounter
reduced, if not entirely eliminated
See if your street will be underwater in the future
some animals can adapt
submerge the homes
suggests the opposite
Summer might be more
that replicates conditions of the pre-existing forest
the drought that hit
the economic impacts of rising temperatures
the plane finally took off
the vintage weather conditions
to even dirtier and more dangerous versions
to overcome the negative impact
try to be hyper-rational about it
we have been studiously ignoring it

"What in the Name of Arse" "Cries out for Universal Brotherhood" "?"

KATRIN BENZLER

"these unnatural men" "barricaded the world with hate"
"crackdowns on" "the love of humanity in"
"endangered species"
"their knuckles and nails black"   "from illegal ivory"
"imprison innocent" "wild elephants"
"who fear the way of human progress"
"these unnatural men" "undoing every protection for"
"the corpse of an elephant"
"hard and unkind"
"and feel too little"
"Human beings are like that."
"And worse than the"
"grindingly dull nature of" "the power to
create machines"
"machine men with machine minds"
"machine hearts!"
"victims of" "Trump"

"You are not machines!"

"A living elephant in Africa is" "more than flesh and blood"

Notes on the Poem

Source texts for this cut-up include:

[1]   The final speech from *The Great Dictator* by Charlie Chaplin

[2]   An article on Trump's decision to lift the ban on elephant trophy imports

[3]   Fragments of a favourite book: *The Tea Rose* by Jennifer Donnelly

[4]   Remnants from the children's book: *Angus, Thongs and Full-Frontal Snogging* by Louise Rennison

Tipped the Scale, Made a List, Then and Now

SOPHIE BERGMEIJER

A variety of steps and thought processes were involved in creating this poem. In researching climate change, ecological problems, and the response to the crisis, I experienced some sense of resistance towards the academic literature. Academic papers usually aim to be factually and statistically correct in transmitting their message, and are often highly theoretical and verbally complex in nature. By using the rational, factual medium of academic literature to educate oneself on the topic of climate change, the emotional aspect of the crisis is often lost. Furthermore, it has been found that people tend to approach the topic of climate change with emotional distance in order to decrease negative emotions such as guilt, fear, anxiety, sadness and anger (Chu and Yang 2019). Despite the factuality and concreteness of academic research, it often unintentionally reinforces the existing distance by its own distanced nature. Art, however, aims to bring messages close.

Poetry plays into people's emotion, despite its frequent lack of concreteness. My first aim was to combine the factuality of academic research with the emotion of poetry to convey the full message; the facts; the concrete, and the emotion; the ambiguous. Therefore the first step of the process was the use of two fragments from academic papers to create a total of three cut-up poems only using the words from these pieces.

Tipped the Scale

Anchored in guilt and shame
vigorous economists agree
economic growth and competition continue
but disregard Indirect effects:
Drought
soil frost
Avalanche risk
wider polarization
the social welfare, preferably taxed
for optimal efficiency

in hope – an abstract emotion
We
engage in empathy
toward victims of climate change
balance
For such a fractious discipline

Made a List

First,
emissions
carbon
plants
drought
largely unaddressed

Second,

manipulation

a matter of politics

starts modestly then

accelerates
anger, anxiety
sadness
fear
perceived psychological distance

resistant

Third,
this study
and empathy
and hope
traverse social distance

Then and Now

rich and poor   snow and ice

                                                      canopy roughness

                                                      photosynthesis

                                                      supplementary

                                                                empathy

                     risk and time

                 online dimensions

                    metabolic rate

                     manipulation

                      resistance

Fire and Fuel
moisture Ignition
people's intention
water deficit
fear and guilt   present and future

According to my grandfather, one day an apple fell from a tree, and left its shape in the clay, inspiring someone to discover the idea of printing. I started to think about things falling, and creating poetry. I decided to make literally fractured poetry. I saw an ad for "seedboms" (seedbombs). The Kabloom company sells compostable paper bundles filled with seeds in the shape of hand grenades. I took three grenades and wrote one of the previous poems on each. I then filled them with water and dropped them in the garden. The words that remained visible became the final poem, though I added words to enhance flow.

When it breaks   First

                                                                    ice

                                                      canopy dimensions

                                                supplementary empathy

                                                                then us

                                Second

                              economists

                          anchored online

                      psychological distance

                              resistance

Third a deficit
fire and fuel
anger
anxiety
sadness
and fear
engage an abstract "we"
                                                      hope only a story

Works Cited

Chu, Haoran, and Janet Z. Yang. "Emotion and the Psychological Distance of Climate Change." *Science Communication,* vol. 41, no. 6, 2019, pp. 761–789., doi:10.1177/1075547019889637.

Seidl, Rupert, et al. "Forest Disturbances under Climate Change." *Nature Climate Change,* vol. 7, no. 6, 2017, pp. 395–402., doi:10.1038/nclimate3303.

Lichen Song and Salt Song

ARTHUR SZE

Lichen Song

—Snow in the air   you've seen a crust on the ceiling wood and never considered how
I gather moisture when you step out of the shower   you don't care that I respire as
you breathe   for years you've washed your face gazed in the mirror shaved combed
your hair rushed out while I who may grow an inch in a thousand years catch the tingling
sunlight   you don't understand how I can dive to a temperature of liquefied gas and
warm back up absorb water start growing again without a scar   I can float numb in
space be hit with cosmic rays then return to earth and warm out of my sleep to respire
again without a hiccup   you come and go while I stay gripped to pine and the sugar
of existence runs through you runs through me you sliver if you just go go go   if you
slowed you could discover that mosquitoes bat their wings six hundred times a second
and before they mate synchronize their wings you could feel how they flicker with
desire   I am flinging your words and if you absorb not blot my song you could learn
you are not alone in pain and grief though you've instilled pain and grief   you can
urge the dare and thrill of bliss if and when you stop to look at a rock at a fence post
but you cough only look yes look at me now because you are blink about to leave—

Salt Song

Zunis make shrines on the way to a lake where I emerge   and Miwoks gather me out of
pools along the Pacific   the cheetah thirsts for me   and when you sprinkle me on rib
eye you have no idea how I balance silence with thunder in crystal   you dream of
butterfly hunting in Madagascar   spelunking through caves echoing with dripping
stalactites   and you don't see how I yearn to shimmer an orange aurora against flame
look at me in your hand   in Egypt I scrubbed the bodies of kings and queens   in
Pakistan I zigzag upward through twenty-six miles of tunnels before drawing my first
breath in sunlight   if you heat a kiln to 2380 degrees and scatter me inside   I vaporize
and bond with clay   in this unseen moment a potter prays because my pattern is out of
his hands   and when I touch your lips   you salivate   and when I dissolve on your
tongue   your hair rises   ozone unlocks   a single stroke of lightning sizzles to earth.

Man to Machine

Sheila karpesky

feathered warbling trees trilling
smoke at dusk fluttering wings
a full new moon opening
bulbous white pupil of new eyes
glimmering with the quiver
new bodies awakening
musing jellied slime-mold minds
pulsing primordial dreams
questioning naming knowing
who are we
starting primitive fires
sopping grease and stabbing meat
stoking the fledgling hunger
first hands scrawling star patterns
carving out our place

withered gasping trees choking
smoke at dawn ruffling dry leaves
a dark moon hanging dying
rain trickles in poisoned streaks
collapsing veins and systems
bodies curling shriveling
rippling at the slightest wind
a deathwatch beetle ticking
clicking counting down the time
why are we
in sudden midday nightfall
blotted sun direful omen
of writhing silvery strands
skeletal hands unworried
digging our own graves

blighted bowing trees sobbing
smoke at midnight stifling heat
ghost orb fading satellite
creeping over wasteland dry
a strange mechanical bird
calling soaring screeching dirge
comet gashing splitting sky
trailing sparks of ice and fire
stardust spilling failing life
what are we
electric singing diode eyes
modified bodies changing
merging man-machine alone
robotic hands void of flesh
overtake our place

Mourning May

CHAD WEIDNER

The world is ending
mainstream science fiction death is preferable to climate
death
word dark to very dark, darker still, dark the, dark garden, the
dark, big eucalyptus trees.

Tin men in tin suits with tin moustaches
autoskeletons, the remnants of humans
assembled wholly or partially from fragments
self-tormenting outcasts, contemptuous of humanity
technological and electronic kingdoms
fancy tv typewriter machines and talking pocket telephones
electric power station vibrates, sparks flying
open mouths with tongues hanging down
frighten the small children.

This is also a fundamental weakness with a certain kind of
postmodernism.
Human beings are now carrying out a large-scale geophysical
experiment of some kind
preemptively making friends with death, and its intimate
companion capitalism.
We cannot say we were not warned
the full implications of burning coal in nineteenth century
England
devouring time,
tiny Pacific islands and the jolly, big-hipped women envel-
oped by the sea
no more birdsong,
our environmental vandalism
totally without meaning.
The old question recurs: America, when will you be angelic?

Sunshine, rain, trees, birds, fish, sky, and prehistoric humans
a wheel turns in the sky.
When I consider every thing that grows
the very future of a habitable planet
stirring, destabilizing, quivering.
Conserve and maintain public natural resources
or apples will sour at your sight.

Hearing the operation of an intensely musical ear
I wriggle in my bed, calling out my lady's name:
Greta Thunberg, enlightened mountain shepherdess
denouncing "fairytales of eternal economic growth!"

They laugh at her youth, intact and whole
and slander leftist liberal elites
and radical humanities professors indoctrinating the youth-
led climate strike.
CEO of Exxon Donald Trump bemoans: "We never really made
any money"
and other impoverished metropolitan strangers
rush off to retail jobs 60% off today only—
cheap Chinese trinkets made by slave labor.

In my head is the picture of all living creatures broken down
to their simplest components of muscles, organs, tissues, and
senses.
The deep inner world.
Oh blessed be the day, the month, the year,
the season and the time.
What would you do if you knew the world was ending?
I'd hug the environmentalists for trying.

Play-back-time
recalling gin-and-tonics, french fries, and deep kissing
clothes scattered on the soil
a bicycle wheel slowly rotates, spokes catching the light
disembodied past
fractured

Notes on the Poem

This cut-up utilizes a number of source texts including bits of
canonical literature, popular articles about climate change
and the economy, postwar poetic fragments, as well as bits
from a writing instruction manual. Lines were selected, modified and mediated, and re-represented. I choose not to be
more specific about the source texts to retain a bit of magic
and mystery.

A Frankenpo:
Report to Exxon Shareholders

KENJI C. LIU

From the President: I am the radiant manager, a hand-pump-
ing diameter of flame ///

California heat
consumes tender area: A major mop-up situation for officials ///

Firefighting crews prepared for boss difficulty
wildfires in dead topography ///

Non-essential living
systems scraped for fertile percentage ///

Database alarmists politicize
life-threatening poor with upward line of dead species ///

Synthetic trees flare-up in customer airspace—
industry officers name Congress extremists ///

Bankrupt
public agency branching its way toward an eight-lane burning
(while 106,000 limbs prune inconclusive science) ///

Petrophobic fish leading meteorological suppression
of corporate oil emissions ///
Local cloud-to-ground activist
attack by generic forest spirits and coyotes diverted ///

Another group darkness/burning timeline:
Interagency oil tankers supply million area stumps with
national flags ///
Brand operations spark atmospheric
salary position ///
Preparedness explosion improves
projected hospital industry profits ///
Ground moisture
resources minimized in favor of emergency soil markets ///
Corrective golf measures and videocassette surveillance
boosted ///
Grab-and-go executive flaming parachute
manufacturing prepares for gusty income genie ////

Notes on the Poem

A frankenpo is an invented poetic text created by collecting, disaggregating, randomizing, rearranging, recombining, erasing, and reanimating one or more chosen bodies of text, for the purpose of divining or revealing new meanings often at odds with the original texts. This frankenpo is a combination of "Energy, the Economy, and the Environment" (a 1996 speech by Exxon Corporation's Chairman to the Economic Club of Detroit), "PG&E: Massive Power Shut-Off to Hit 800,000 Customers, Could Extend Nearly a Week" (*San Francisco Chronicle*, October 8, 2019), "After Choosing Profits Over Maintenance, California Utility Giant Forces Blackouts on Customers" (*Salon*, October 19, 2019), and a glossary of "Fire Terminology" (Glacier National Park website).

Pests

ANDREW SHAW

Lost places where skulls meet

*Where do they come from, those creepers and crawlers that sud-
denly appear on house plants much to the dismay of the owners?
Plant pests come from many sources.*

bargain basement banter below

*the best way to prevent insects from colonizing plants is to give
the plants good regular care. Frequent misting and washing of
foliage,*

the body itself, a shelf

*helps keep plants pests away. A new plant should be isolated for
several weeks before it is added to a major collection.*

What's the right thing to do?

Many pests can be controlled by taking plants to a sink where the foliage can be washed under the faucet.

the collapse of Western civilization

For heavier infestations, spraying may be necessary. Spray in aerosol cans clearly marked

eat healthier foods. For lunch, barbecue chicken. Yes, sir. Ape mind in full effect

The insects most likely to be found on house plants are described below.

Live in another dimension, if you like. Free to leave at any time

House plants infested with aphids, mealy-bugs, or scale often attract ants because these pests excrete a honeydew

Funeral scheduled today for President George H.W. Bush

Where they suck the plant juices and cause stunting as well as distorted buds and leaves.

Miraculous spelling

Some kinds feed on roots

Be seated, soldiers.

The burrowing in and around roots may do more harm than good, and the value of castings is minimal in the small confines of the pot.

Expressing what's inside. Taking deep breaths

The maggots are most prevalent in soil mixtures containing high quantities of humus, pest moss, or leaf mold.

To hereby erase all politically motivated posts

These insects are covered with a white fuzz that make them look like specks of cotton.

It is indeed good to express dissatisfaction.

The pests have favorite feeding places: in leaf axils, on the underside of leaves, along leaf veins, and inside the protective coverings of flower buds.

I hope I can be forgiven.

For a heavy infestation, isolate the plant and dip it in a large
bucket of melathion liquid. Mix this by putting two teaspoons of
57 percent emulsifiable concentrate in each gallon of water.
I don't know where to start.

More a nuisance than a serious pest,

I'll write a note

Unless controlled, red spider mites can destroy plants.

The evening was scattered.

Notes on the Poem

Italicized excerpts taken from: Joan Lee Faust. *The New York*
Times Book of House Plants. Quadrangle Books, 1973.

FRAMING DEVICE:
The Animal's Perception of Earth

EDWIN TORRES

[what forms you, what humans you]

In what form will the scattered emblems of our ecology be most clearly received – to consider all the colors of our arrival as animals, all monolithic immersions that equate earth to body? In the fragments of our human communication that desire arrival – receptive ecologies that conform the human condition to human nature – where can language assume fracture, to frame change as the inevitable passage, perceived, by the inevitable human?

[what opposes you, what bends for you]

Are you willing to leave your life – as you know it for the chance to change someone else's?

Where have you placed yourself – to be viewed by others
to be invisible?
   Thrust in a revolution – by the stance of my strut
: who can   change the world
                   nada
: what can   change the world
                   el viento
: cross currents   strike across
                   the globe a flutter
: quakes

the passage of time is shattered... by elements of nature
      the life of man is shattered... by that of another
      man is nature to man... nature is nature to nature

[what I have for you, what arrived for you]

   To consider earth as arrival, for words to gravitate towards
– to settle on timelessly gorgeous derelicts of language – to
settle on the floatation devices, wrapped around, the layered actions that define change – what will our words do, to
change the earth – with the places we visit, with language,
with considerations of placing, with how we own our landings – what will the motivation of ending, do – to change the
ending, to land or maybe, only, arrive – in action of supple
ability, to the human body – not compliant, not flexible, not
allowed until told so, the human mind, is a place inside earth,
told to be earth — to be the human, telling the mind

to arrive — the human arrival, not pliant until given words to
assemble around, to land —
with gentleness, the force needed for change.

To strike familiarity, in consideration of placement — of
earth as a place, in consideration of alphabet — to allow inte-
gers of cryptic consideration, their translation, for creatures
of cognition — to absorb intention, to allow, of place — the
human earth, its roam — for giving place, what we do with
time — inside earth, to mine through human change, consideration for change — to note, how long it takes to set out,
what words do — to ignite the human, out of the human — to
arrive as a fuse — to note the implosion of language, and how
it affects our human mind — to consider the mind, as arrival,
to allow layered action, its depth, complicity to depth — to ar-
range notions of human complexity, as mistake-filled *prones,*
to own our *prones* to action — to be prone to action and
assemble abilities, as triggers — to use language that implies
friction, as a need for change — to know that change is action
out of reversal, out of what triggers implosion — to ignite
evolution, by writing in ways, that don't arrive easy.

Unaccustomed to being caught — the human mind is not
earth, is not alphabet is not a stranger to words, as a place for
landing — unaccustomed to stagnant episodes of hierarchy,
the human earth is not a stranger to evolution, to the re-done
resonance of false implications — the human earth, in mind —
to the mirrored imperfections of the rendered obsolete, the
human earth, in mind — to the chosen arrival, a disguise of
arrival – to the rusted ladder, that blends with the wall, to the
manufactured escape of the human spirit, the construction

of place, as consideration, for landing – to apprehension as
sport, for the human mind – to barriers invented, unaccustomed to spirit – to words that absorb intention, by reframing
ability, as punctuation liberators – of earth as liberty, in face
of spirit, to allow human arrival its place – by reframing place,
to gravitate towards change, by reframing change – to set out
timelessly, the human earth, as a mind – in consideration of
arrival.

[what hands you, what thumbs you]

    The elements have — their own pecking order
just like humankind.
    : what person
can change the world?
    : is one
in tune with nature?

[what isolates you, what powers you]

    we find ways to isolate truth
    to isolate power
    by calling on the shapes that save us
    the shapes that survive us
    out of brain into body

    we call on our instincts
    to isolate hurt
    away from emotion

to protect our power
we call on our truth
to attempt deflection out of heart

we call on our hurt
to isolate deflection
away from the shapes that save us
the ones that define us
are the ones we call on
to survive our truth into hurt

we call on our finding
out of knowing
away from doing
we find ways
to isolate truth

[what frames you, what knows you]

Inside the unknowable new lives I've never met
the ones I've yet to understand, are those who live in shadow
and told to avoid shadow.
          Those who live in sunlight
and told to avoid giving.
          Within the seeing ones, who only now
          I've never met, are those
          who I never intended to disappoint
          — as I didn't know
          how to take or give or say, until now.

[*the closer I get to the echo, the closer I hear*
[*how many times I missed the mark*

The only ways to say you change, are to change the ways
you say them. Fragments swirled in the never-ending
listening that repeats your name.
                    Those who say your name
or at least know your end,
                    before you knew to end your knowing,
are those who repeat
                    your dimension – to end
                    your dimension, the one you stand in.

[*we envision the dimension released,*
           as the size imagined
[*back when our size knew who we were –*
           back
[*when our blessed fragments appeared*

Play your size on repeat. Hear the loop, to hear
something new – as holding as holding as holding – was
there a soul you wanted to mate with?
                    A half to who you were?
A repeating intelligence
                    that you left back – to never be?

[*the compartment of your depth, a volume to*
           your sunlight

[*has that been unreleased – until you hear*
               again

[*of one more life you didn't know? a frame*
               you became?

[*a constructed opening, closer to the one*
               you're in?

Don't give away the moments you had, before you
knew that time would relentlessly
        obsess over change. The cracked impossible
you wanted to manufacture
        out of heartless abilities – for intertwined
arousal.

[*a fragment of your life reflected – by the*
               parallel divinity

[*you search for – the one you avoid is the*
               one you never leave –

[*but again, to your cornerstone, the one we*
               started with

[*the reach you never knew was possible,*
               until now

[*the reach of now – the reckoning of that*
               loop

The mention of a loop, on repeat — the facing of
time.   The hearing of me
        saying, what words, you think you say.

[*to live in abundance – and told to avoid ending*
[*to live in reflection – and told of what stops you*
[*of who meets you, to avoid the moments you*
              end
[*to live the hearing – of the moment, on a loop –*
              the hearing
[*of a loop, without a frame – here now*
[*the moment of your name*

[what animals you, what wants you]

I had animals lined up... to smell yours...
I once had similar creatures
but I was a different tone...
I once put strokes to stride... but I was a different
sort of dog... I once had gills at the side
of my testaments... but I was
a brilliant miscreant... I once had scent
as an available orifice
but it was someone else's skin... I let fall...
someone else's ocean
I let in... someone else I could not imagine

I once wanted so badly
to feel everything – that I let go
of the now – to let it happen

I once knew the travel of my sweat
would unleash a tremor

so far beyond my sight — that I held onto perception
beyond my skin — to let it happen

I once needed affirmation's reminder
with such feral acuity — that I let go
of my size — to let it happen

I once thought ferocity wanted me
for mobility
with centers revolving —
that I smelled the spot
at the scratch of
my back — to let it happen

I once allowed myself the walking through
of an air so dense with temporal serenity — that I
held on to registers
incomprehensible to acute attenuations
of interlingal
arousal — to let it happen

I once telegraphed my tremor to the sacrum
before the cortex
with such badass delivery — that I let go of a
torment I'm only now
understanding was deeply seeded in my
twinned removal
of seed and sift — to let it happen

I once wanted to know
            whether I could attract the subtlest domains
                    of sensory arousal — that I stoned my hide

entranced by delineation
the animal perceives – and lets it happen

[what mediates you, immediates you]

        Are you willing to leave your life – as you know it
                    for the chance to change someone else's?
            Where have you placed yourself – to be viewed by
                                                                    others?

[what serves you, what saves you]

It is in *fracture* that the emblems of humanity set root,
where the initial stages of cognition assemble their nerve
endings, to create, from that which was left over. The sem-
blance of thought, in its origin as flesh, is manifest by the
reception of flesh – to envision fragment as a whole, is to
embody fracture as immersion. This is language taking root
in fracture. The form of that fracture is what poetry allows
for – to raise questions among the detritus of alphabets that
humans survive on. If poetry is an animal in the landscape of
our bodies, the question of the animal is to survive our truth,
to call on our fractured forms to save us.

On Cures & Abrasions
from a Responsive Environment

ANGELA PEÑAREDONDO

what was once wild   chicken & pig farms   once hectares
of pineapple fields
banana coffee cacao rice corn now inedible impass-
able uninhabitable
from village to village she tramples   her land extends
beyond parameters
hijacked   by globalization   process known as ra-
cial&cultural cleansing

families&homes washed by volcanic dust   it rains like this
in abundance
into her briny tinctures   futurity & ashfall bond in her
throat   concepts

from the survivor   blast after blast   chopped burning portions of earth

shelter an architecture   of fallen switches   from her
hands herbs she rinses
in Taal's ruined lake   if not from the violence of water
then fire
"Capitalism, having already found a way to turn profit on
disaster,"
Aimee Bahng writes on "homeland futurity"

she formulates   teleportation devices   out of what
empire identifies as poor   as root bark   as docile
does surveillance mean the same as command
                            remember what it's like
        to be   of this deserving place   she wipes
down the fevered
deemed unproductive   makes what's bruised
aromatic

because this is what   she has   technology against
incendiary epistemes
she continues   to scan for any of the living birds
                                        through the barrel of a
sugarcane stalk

Survivor's Topography

like an upward buttress a guar skull & her horns calls in evidence outside a catholic church
skeleton :: coral of Aklan :: contractual object

from the dead horizontal & hanging—ask what signage occur
as schemata for rescue   whose oars that lifted a boy out of
floods its plains a site of the anticolonial experience

in direction of what now possesses you   i pray at 2.5
meters above sea level   my child   a burning celestial
reverence for labor expands with untranslated vocabulary

contour through brown scales of this flesh   union of mountains   confusion of phragmites   a divine missive   water
:: measure for haunting   what need   needs not
saving

before my birth rivulets   wonderment of bones splayed
out like numerology across an eroded alter   purge childhood trauma an eruption   bio social
geo me   try

solitude & work merge within you   i offer a bed of moss &
teak bark from this submersion   somewhere in these mili-
tarized lands   there are rivers   there are nets

there are   before my birth there are   bodies you can
no longer see
                  from this excavated   dream   map   made
&   unmade

Exigencies in Layers I

sediment [7]: abuser, attacker, interrogator, ungrateful &
exorbitant   you do not end with age

sediment [6]: to vanquish   one's crucial identity

partial lithification [5]: how could I be a much more divine
being beyond my shame? tell me what it means to view this
body outside of a fetishized object or extinction

sedimentary rock [4]: mattresses split in two

shale [3]: when your body is not your own & only exists out of
the desire from others

there is too many of us out there

sedimentary conglomerate [2]: It feels as if you are farther
from me now, even as I flee or fight off flashes, you simply
desire my awry

breccia [1]: forgive us even in our deviancy, fossil of hope

Exigencies in Layers II

cuticle [1]: when one thinks of justice what worlds do you
inhabit can you be your own mirror your naked privilege the
earth floor in shadow

upper epidermis [2]: how do you transform others if you are
in need
of that brutal & necessary act   who is your realized self
when hurt or hunted

epidermal hair [3]: birdsong renders to tiny explosions   in
the veins of the pacific archipelagos

substomatal chamber [4]: if you want to transform me you
must sacrifice with me

palisade mesophyll [5]: when was the last time, mother
you loved yourself   before others
before you left   before you washed & scrubbed the back-
sides of these monolithic   structures that
collapse   your vascular   bundle
sheath   mother   we ask perpetually
in disaster

xylem [6]: it should have been you father ?
father these sediments   father droplets of giving
mist   father winged & wingless   father
wanted   father to notice us   father failure
father cause   father surface   father be
part of sunlight
   or decay

air channel [7]: *blasts evolve a range of*   *the shock*
is no longer in me

guard cells [8]: collective praying hands

stoma (pore) [9]: not out of essentialism or the failed notions
of universalist matter
i exist   to exhibit the imperfections of men   of the
church   of those violent
        senescent pathologies that assume the form of tradition

air channel [10]: ask me what I want

phloem [11]: ask me what I need

chloroplasts [12]: what's your method of asking

lower epidermis [13]: the invisible   the ghosts   the
disabled
who able word & ash   unremit-
ting   counter   meanings

Why Grandfather Turns Manatee

no data reported of your kind here not in these lowlands too
wet & too flooded termite

mounds grow to mausoleums earth by muddy earth humidity & unicode of incorporeal

transcribe jungle paradigmatic struggle seagrass weep integrity of your birth

your body sank to sexual economies when soldiers strangled
your lover before using

them up first in a burned out salt barn damn military thought
them comfort woman

before they figured our queer lineage without worship of
brown archipelago femmes

you ask your dead lover the same you ask yourself what feminine did supremacy

assassinate you decide to turn manatee like a feminist
un-fracturing conquest

of the feminine you ache to live where border crossing is
your biology proclamation

you were made to partner in deep softness & ancestor undoing without condition

full:new

JESS ALLEN AND BRONWYN PREECE

my body understands niche.branch through bone.time chronicle(r): decaying leaves, warrior lichen
raven's call enters pelvic vessel, burroughs east/west: rounded somatic metronome . feel greeted
noon's marionette.suspended.tide's puppet.what is reflection? sharing gravity.between breaths
site of conception   --   ovaries full, but not emptying   --   join mating birds
vertebrae trunk, my undulating spine: sipping sap: lubricating xylem prayer
always each other's silhouette: in constant contact, contacting constantly
cheek impacts cliff: falling fallen: a seismic viewpoint
soleprints : (im)prints of soul in snow : softness
beyond before answers questions : sounder/mover
limbs entwined, weighted torso tangle
sky chanter melting symphonics
symbiotic curvatures [dis]appearing
sunset's zygote
wombed
bleed
calon lân
blue deity humming
silver sliver, adrenalin flight
swallowing anxiety in stillness, saliva
licking skin to retching, vacant-vernal uterus
slap of white surf sounds between shoulder-blades
rock yields a soft baptism: red dripping onomatopoiea
resolving place, finding niche: warm like liquid mineral veins
weather in my bones, this handful of collaged pebble phalanges
being still, still being; alert to sonic silhouette of others unwelcome
sedimentary scribble: grinding gut to rock, revealing the taste of time
shelved on rock with tight heart while waves are a whale blowing upwards
singing chromatic scales sound stratifies, while, in greyness, colour is rupture and rock bleeds

poisonous pendulum swinging, the altar of my palm, mermaid feet buried carefully like eggs
making a nest of limbs, clinging on for biological-life out of my geological-depth
outstaring me the vacuum-darkness, time deeper than I am allowed to know
photosynthetic urchin crab my mouth open in my naval, sp(l)itting sun
slithering, shock of bird's sharp call slamming into solar plexus
saline fluid reflecting the angry mineral darkness inside me
watching not fighting; pressure on edges of emptiness
crevice-wedged, I'm curious at softness of self
giving voice to a dangerous wildness
synovial jiggle, raving on rock
trust this falling duet
sky ripped open
under shell
sigh
elemental
precipital release
sensuous ancestor encounters
shamanic opening : recapitulating spirit
metaphor woman : liminal frog relation
unfurl(ed) shoulder(ed) wings: herstory takes flight
processual journey of understanding: deep relinquishing, shedding
fog is my shadow: saturated presence alludes me
nearing equinox my new shoots slowly push upward, rising
i vocalize land's tongues, strung/sung on night's heralding call
the 'and' in morning's call-*and*-response : noticing avian soundscape mapping
deer (dear) meetings : intraspecies communication – pivot point – photosynthetic relay, lung to
pre-dawn's amphibian chorus fills dark contours: my heart cavity tethers me home
light streams my mossied hair – soaring eagle antenna -- fullness reveals 'my' 'own' synesthetic
dichotomies....

poisonous pendulum swinging, the altar of my palm, mermaid feet buried carefully like eggs
my body understands niche.branch through bone.time chronicle(r): decaying leaves, warrior lichen
making a nest of limbs, clinging on for biological-life out of my geological-depth
raven's call enters pelvic vessel, burroughs east/west: rounded somatic metronome . feel greeted
outstaring me the vacuum-darkness, time deeper than I am allowed to know
moon's marionette.suspended.tide's puppet.what is reflection? sharing gravity.between breaths
photosynthetic urchin crab my mouth open in my naval, sp(l)itting sun
site of conception   —   ovaries full, but not emptying   —   join mating birds
slithering, shock of bird's sharp call slamming into solar plexus
vertebrae trunk, my undulating spine: sipping sap: lubricating xylem prayer
saline fluid reflecting the angry mineral darkness inside me
always each other's silhouette: in constant contact, contacting constantly
watching not fighting; pressure on edges of emptiness
cheek impacts cliff: falling fallen: a seismic viewpoint
crevice-wedged, I'm curious at softness of self
soleprints : (im)prints of soul in snow : softness
giving voice to a dangerous wildness
beyond before answers questions : sounder/mover
synovial jiggle, raving on rock
limbs entwined, weighted torso tangle
trust this falling duet
sky chanter melting symphonics
sky ripped open
symbiotic curvatures [dis]appearing
under shell
sunset's zygote
sigh
wombed

elemental

bleed
precipital release
calon lân[1]
sensuous ancestor encounters
blue deity humming
shamanic opening : recapitulating spirit
silver sliver, adrenalin flight
metaphor woman : liminal frog relation
swallowing anxiety in stillness, saliva
unfurl(ed) shoulder(ed) wings: herstory takes flight
licking skin to retching, vacant-vernal uterus
processual journey of understanding: deep relinquishing, shedding
slap of white surf sounds between shoulder-blades
fog is my shadow: saturated presence alludes me
rock yields a soft baptism: red dripping onomatopoeia
nearing equinox my new shoots slowly push upward, rising
resolving place, finding niche: warm like liquid mineral veins
i vocalize land's tongues, strung/sung on night's heralding call
weather in my bones, this handful of collaged pebble phalanges
the 'and' in morning's call-and-response : noticing avian soundscape mapping
being still, still being; alert to sonic silhouette of others unwelcome
deer (dear) meetings : intraspecies communication – pivot point – photosynthetic relay, lung to leaf
sedimentary scribble: grinding gut to rock, revealing the taste of time
pre-dawn's amphibian chorus fills dark contours: my heart cavity tethers me home
shelved on rock with tight heart while waves are a whale blowing upwards
light streams my mossied hair – soaring eagle antenna -- fullness reveals 'my' 'own' synesthetic dichotomies....
singing chromatic scales sound stratifies, while, in greyness, colour is rupture and rock bleeds

---

[1] Welsh: "pure heart"

Composition Statement

> This, then, is Uncivilised writing. Human, inhuman, stoic and entirely natural... Apart, but engaged, its practitioners always willing to get their hands dirty; aware, in fact, that dirt is essential; that keyboards should be tapped by those with soil under their fingernails and wilderness in their heads.
>
> — Hine and Kingsnorth

"full:new" was a transnational experiment in embodied writing over a lunar cycle in February and March 2014, performed by two ecological artists/scholars, situated on two different continents and two different countries (Wales/Canada). It was a sensuous layering of enquiry into what emerges from the commitment to write in response to moving daily on a chosen site: an exploration of the relations that exist between touching the soil and soiling the paper, bodies in the world and fingers on the keyboard. By moving, we are referring to what could usefully be categorised as site-specific dance, which has multiple definitions and aesthetics, but in our joint practice, the focus lies in a literal dissection of the term *kinaesthetic:* kin as implying animate relationality with world, and aesthetic as a conscious performance process, but one not dependent on the presence of a human audience.

Beginning on the full moon, we decided we would each visit a selected site daily for 28 days, each day observing a moving-writing score for 28 minutes. Our ultimate aim was to produce a collaborative and concrete poem that captured in its structure and its visual form the changing shape of the lunar cycle: the figurative appearance, disappearance and re-appearance of the moon. The score for writing followed the course of the lunar cycle in that each day's moving gave rise to a single line, the word length of that line dictated by the wax and wane of the moon. We began on the full moon with a line length of fourteen words, decreasing daily by a single word, arriving at one word on the new moon. At this mid-way point we shared our halves with each other for the first time, then over the final fourteen days waxed back our line length from one to fourteen words by the next full moon.

The "full:new" poem exists in a format that best reflects, to us, form, narrative and process. The poem is a reflection of the moon (and its circularity), the two halves of text mirrored on both sides of the page. The lines of each author are not in-terleaved and each quarter remains intact though they are ex-changed in such a way that the layout reflects the process of our sharing of quarters at the half way point.

The full:new poem was written line by line each day in a kind of interleaving (one Allen line, followed by one Preece line and repeated). A different narrative inevitably emerges as our sites and personal histories meet across different hemi-spheres. It was immediately apparent to us that colloquial-isms of region, locale and habitat were determined to emerge in our own writing, informing the verse in both expected and unexpected ways. The sites were satisfyingly contrasting: Al-len's a shingle beach in Cardigan Bay, Wales beneath cliffs of turbidite deposits, largely devoid of vegetation; Preece's a moss-covered rocky bluff overlooking the Salish Sea in the Coastal Douglas-Fir ecosystem of Canada. The vernacular was personalised to the extent that the geological, geographical and biological features became inextricable from autobiographical expression.

A layout emerged – one which owes its form to a fortuitous error of display in an iPhone email program as we were exchanging alternative arrangements of lines. While the original had not been arranged thus, it was the way the screen chose to display the poem, thus introducing a technological chance element to the composition process. To us, this arrangement speaks of some kind of birthing, redolent of fertility, movement, and so was recreated and retained.

While our language often acknowledged the presence of 'external others' we felt/feel that our naming of things does not automatically imply a separation from them. Though we were equally conscious of the challenge of not objectifying nature through lauding it, falling prey to the lure of Romanticism, with its construction of an externalized 'wilderness' (Morton). Rather, our language continuously sought to contextualise a whole: a holistic weaving-in-place. Performance-writing practices such as full:new which explore the liminal zones between bodies and between states – of moving, writing, being – just may be placed to capture the subtle seismic pulses of what resides elusively on the periphery of our (human) awarenesses...

Works Cited

Hine, Dougald and Paul Kingsnorth. "Uncivilisation: The Dark Mountain Manifesto." The Dark Mountain Project, 2009.

Morton, Timothy. *The Ecological Thought. Harvard* UP, 2010.

Tender

ROBYN MAREE PICKENS

The design of the mitigation portfolios and policy instruments to limit warming to 1.5°C will largely determine the overall synergies and trade-offs between mitigation and sustainable development (very high confidence). Redistributive policies that shield the poor and vulnerable can resolve trade-offs for a range of SDGs (medium evidence, high agreement). Individual mitigation options are associated with both positive and negative interactions with the SDGs (very high confidence). {5.4.1} However, appropriate choices across the mitigation portfolio can help to maximize positive side effects while minimizing negative side effects (high confidence). {5.4.2, 5.5.2} Investment needs for complementary policies resolving trade-offs with a range of SDGs are only a small fraction of the overall mitigation investments in 1.5°C pathways (medium evidence, high agreement). {5.4.2, Figure 5.4} Integration of mitigation with adaptation and sustainable development compatible with 1.5°C warming requires a systems perspective (high confidence). {5.4.2, 5.5.2} (Roy et al. 448).

T/t

Tender design of tender mitigation portfolios and policy instruments tender limit warming tender 1.5°C will largely determine tender overall synergies and tender-offs between mitigation and sustainable development (very high confidence). Redistributive policies tender shield tender poor and vulnerable can resolve tender-offs for a range of SDGs (medium evidence, high agreement). Individual mitigation options are associated with both positive and negative interactions with tender SDGs (very high confidence). {5.4.1} However, appropriate choices across tender mitigation portfolio can help tender maximize positive side effects while minimizing negative side effects (high confidence). {5.4.2, 5.5.2} Investment needs for complementary policies resolving tender-offs with a range of SDGs are only a small fraction of tender overall mitigation investments in 1.5°C pathways (medium evidence, high agreement). {5.4.2, Figure 5.4} Integration of mitigation with adaptation and sustainable development compatible with 1.5°C warming requires a systems perspective (high confidence). {5.4.2, 5.5.2}

E/e (1/2)

Tender design of tender mitigation portfolios and policy instruments tender limit warming tender 1.5°C will largely determine tender overall synergies and tender-offs between mitigation and sustainable development (very high confidence). Redistributive policies tender shield tender poor and vulnerable can resolve tender-offs for a range of SDGs (medium tender, high agreement). Individual mitigation options are associated with both positive and negative interactions with tender SDGs (very high confidence). {5.4.1} However, appropriate choices across tender mitigation portfolio can help tender maximize positive side effects while minimizing negative side tender (high confidence). {5.4.2, 5.5.2} Investment needs for complementary policies resolving tender-offs with a range of SDGs are only a small fraction of tender overall mitigation investments in 1.5°C pathways (medium evidence, high agreement). {5.4.2, Figure 5.4} Integration of mitigation with adaptation and sustainable development compatible with 1.5°C warming requires a systems perspective (high confidence). {5.4.2, 5.5.2}

N/n

Tender design of tender mitigation portfolios and policy instruments tender limit warming tender 1.5°C will largely determine tender overall synergies and tender-offs between mitigation and sustainable development (very high confidence). Redistributive policies tender shield tender poor and vulnerable can resolve tender-offs for a range of SDGs (medium tender, high agreement). Individual mitigation options are associated with both positive and tender interactions with tender SDGs (very high confidence). {5.4.1} However, appropriate choices across tender mitigation portfolio can help tender maximize positive side effects while minimizing tender side tender (high confidence). {5.4.2, 5.5.2} Investment tender for complementary policies resolving tender-offs with a range of SDGs are only a small fraction of tender overall mitigation investments in 1.5°C pathways (medium evidence, high agreement). {5.4.2, Figure 5.4} Integration of mitigation with adaptation and sustainable development compatible with 1.5°C warming requires a systems perspective (high confidence). {5.4.2, 5.5.2}

D/d

Tender tender of tender mitigation portfolios and policy instruments tender limit warming tender 1.5°C will largely tender tender overall synergies and tender-offs between mitigation and sustainable tender (very high confidence). Redistributive policies tender shield tender poor and vulnerable can resolve tender-offs for a range of STGs (medium tender, high agreement). Individual mitigation options are associated with both positive and tender interactions with tender STGs (very high confidence). {5.4.1} However, appropriate choices across tender mitigation portfolio can help tender maximize positive side effects while minimizing tender side tender (high confidence). {5.4.2, 5.5.2} Investment tender for complementary policies resolving tender-offs with a range of STGs are only a small fraction of tender overall mitigation investments in 1.5°C pathways (medium evidence, high agreement). {5.4.2, Figure 5.4} Integration of mitigation with adaptation and sustainable tender compatible with 1.5°C warming requires a systems perspective (high confidence). {5.4.2, 5.5.2}

E/e (2/2)

Tender tender of tender mitigation portfolios and policy instruments tender limit warming tender 1.5°C will largely tender tender overall synergies and tender-offs between mitigation and sustainable tender (very high confidence). Redistributive policies tender shield tender poor and vulnerable can resolve tender-offs for a range of STGs (medium tender, high agreement). Individual mitigation options are associated with both positive and tender interactions with tender STGs (very high confidence). {5.4.1} However, appropriate choices across tender mitigation portfolio can help tender maximize positive side tender while minimizing tender side tender (high confidence). {5.4.2, 5.5.2} Investment tender for complementary policies resolving tender-offs with a range of STGs are only a small fraction of tender overall mitigation investments in 1.5°C pathways (medium tender, high agreement). {5.4.2, Figure 5.4} Integration of mitigation with adaptation and sustainable tender compatible with 1.5°C warming requires a systems perspective (high confidence). {5.4.2, 5.5.2}

R/r

Tender tender of tender mitigation portfolios and policy instruments tender limit warming tender 1.5°C will large-ly tender tender overall synergies and tender-offs between mitigation and sustainable tender (very high confidence). Tender policies tender shield tender poor and vulnerable can tender tender-offs for a range of STGs (medium tender, high agreement). Individual mitigation options are associated with both positive and tender interactions with tender STGs (very high confidence). {5.4.1} However, appropriate choices across tender mitigation portfolio can help tender maximize pos-itive side tender while minimizing tender side tender (high confidence). {5.4.2, 5.5.2} Investment tender for complemen-tary policies tender tender-offs with a range of STGs are only a small fraction of tender overall mitigation investments in 1.5°C pathways (medium tender, high agreement). {5.4.2, Figure 5.4} Integration of mitigation with adaptation and sustainable tender compatible with 1.5°C warming tender a systems per-spective (high confidence). {5.4.2, 5.5.2}

Works Cited

Roy, Joyashree, et al. "Sustainable Development, Poverty Eradication and Reducing Inequalities." *Special Report: Global Warming of 1.5°C,* Intergovernmental Panel on Climate Change, 2018.

Whose Who

A RAWLINGS

Who owns the environment? Who owns wolves? Who owns owls? Who owns turtles? Who owns moss? Who owns moose? Who owns shrews? Who owns flowers? Who owns herons? Who owns trout? Who owns trees? Who owns selves? Who owns flies? Who owns moths? Who owns mushrooms? Who owns vultures? Who owns noise? Who owns letters? Who owns love? Who owns the north? Who owns the view? Who owns where? Who owns south? Who owns west? Who owns elsewhere who owns nowhere Who owns territories who owns terror who owns Enemies who own loss Who owns winter who owns summer who owns eyes feet hoots hooves Who owns less? Who owns shores Who owns snow Who owns nest or fruit Who owns hormone levels Who owns mothers who own forests who own life who own mouths, teeth, volume Who owns new Who owns the litter Who owns synonymy Who owns howl Who owns three to five months Who owns one (only), two (only) Who owns wherever Who owns flesh owns soil Who owns hollow trees Who owns sense of smell Who owns eerthworms, voles, snwils Who owns North Flmeriflfl Who owns flests for the flites who own the environment Who owns the environment Who owns the environment? Who owns my reverie? Who owns my stone? Who owns my moss? Who owns my fever? Who owns your fewer? Who owns her turtle? Who owns his moose? Who owns this mist? Who owns their must? Who owns whose howls? Who owns whose owls? Who owns how now? Who owns whom? Who owns who? Who? Who?

Notes on the Poem

This piece is part of a larger project called *Echolology*, which treats the book as a closed linguistic ecosystem while interrogating and indicting the English language's communicative reliance on pronouns and their inherent hierarchy. *Echolology* performs an ecopraxis, incorporating inherited procedures and poetic forms (lipogram, cut-up technique, homophonic translation, erasure) co-opted to emphasize a green ethos in writing practice through reduction, recyclage, and reuse of language. Lipogrammatic constraint is present in the excerpt I provide, as the poem is written, as all poems are written using only letters (e f h i l m n o r s t u v w y) found in pronouns (I you he etc.). Erasure technique is also present in this excerpt.

Feral Poetics for a
(Re)Programmed TechnéPolis

MARI-LOU ROWLEY

A Random Walk Home and Back Again

As someone who has lived most of my adult life in large Canadian cities of Edmonton, Toronto, and Vancouver, I found moving back to Saskatchewan an unexpectedly alien experience; I felt like an outsider in my own hometown. As a writer whose work has been influenced by urban avant-garde literature, visual art, performance, and dance, as well as science, I found my poetry regarded with suspicion by a community that fosters and champions "prairie realism." Yet I drafted most of my books at writer/artist colonies in the province, and despite the winters, it remains my favourite landscape in the world for its vastness and wide-eyed unblinking sky. And as Anne Szumigalski, my first poetry mentor, quipped, it is a place where the weather can kill you any time of the year if you are not prepared.

Anne, with her English background, wealth of knowledge, and boundless creative imagination, was a stubbornly original and "disobedient" poet. She would have agreed with Alice Notley, who said disobedience is "predicated on leaving in as much mind fuzz as possible, that is being open to all that is out there in all telepathy – not a very organizable entity, the entity." Anne, too, "allow[ed] in notions from dreams... odd images to take on the weight of truth; and [was] stubbornly involved again in what you might call mystical conceptions" (Notley). Anne was a Blakean who believed in angels, after all. She was not interested in the stark, folkloric narrative of prairie realism; she embraced – and imprinted on me – her own prairie surreal poetics, as in this poem from *Suicide Psalms.*

DAY FOURTEEN: SNOW SIRENS

for Anne Szumigalski

Gaze-dazzled
snow showing off
a spectroscopic dance
    blue green red
frequencies so clear and pure
you want to remember
only this.

Waist-high drifts   beckon
for the crush of crystals under hushed footsteps
on windblown spines.

Let us throw our snowy limbs around you
here, lie down with me
my angel
spread your arms/legs
wide to the sky   feel
the gentle tug of earth,
the caress and release of sinking.

Close your eyes.
Feel your breath crystallize and rise
one last time. (Rowley 83)

As I revisit and revise this essay, it is –35 Celsius outside, –47 with the windchill, which at this extreme cold is roughly the same in Fahrenheit. My outdoor thermometer bottoms out at –50 Celsius and –60 Fahrenheit. Here on the Saskatchewan prairie, even with global warming, winter can be so harsh that I wonder how birds and animals survive outdoors. Before the cold snap, I deliberately put out special feed for the chickadees and nuthatches that flitter about my treed yard, but only a few have ventured down from their sheltered perches high in the spruce and elms. I think about how the frigid cold and darkness challenge survival on all levels: physical, emotional, spiritual, mechanical (cars not starting, furnaces failing, locks freezing shut). Something animal, something feral triggers survival mode. I gas up the car. Stock up on groceries. Put in a new furnace filter. Get out my grandmother's fur coat and wear it with impunity; the fur makes me feel invincible, powerful, beautiful, wild. But there is a darker and deeper side to the emotional and physical effects of this weather, this climate, this place, a hibernation, or worse, a gentle lulling to sleep after the freezing air numbs pain and hypothermia has set in. Do animals feel this too? I wonder. Is that why the chickadees and nuthatches are not venturing forth for food? Is that why my cats want to sleep all day? How temperature and barometric pressure affect the mind-body. Yes, this weather can kill. Take shelter or die.

As colleague, poet, and English professor Hilary Clark reminded me, Anne was the original feral muse in this province. As Hilary writes, quoting a poem of Anne's: "Lock up the poem in too firm a purpose, and like a child, it 'leans from [its] window listening for animals/ far away in the woods,' plotting escape or surprise" (197).

Saskatchewan is a place where, as Pamela Banting stated in her plenary talk "Making Scents: Signature, Text, Habitat," you need the ability to "read for your life." Indeed, since moving back I have come between a black bear and her cub, encountered a timber wolf, learned how to catch and fillet fish, recognize cougar tracks and bear scat, and shoot a rifle. These experiences have made their way into my poetry as well, partly as content, but more importantly as praxis. Yet I am aware that my lived experience of this place, my "intentionality" in the phenomenological sense, is shaped by my urban self as well. Feral poetics comes out of this struggle of reconciliation and reconnection with my poetic roots, a re-wilding.

Other poets have thought and wrote about the wild, the animal, in poetry. I met Dorothea Lasky in New York in 2008. I was in the city launching Suicide Psalms and while there attended a reading of Ugly Duckling Press authors. I was an

annual subscriber to the press and was inspired by Dorothea's UDP pamphlet *Poetry Is Not a Project,* with her simple
yet poignant message – blatantly nonacademic coming from
a poet who teaches in the academy. She writes, "Real poetry
is a party, a wild party, a party where anything might happen.
A party from which you may never return home. Poetry has
everything to do with existing in the realm of uncertainty." Tonight, when I was searching for the pamphlet, which I thought
was among my stacks of chapbooks, I came across *Water Marks*
by Keith Waldrop, *Laments* by Jenny Holzer, *Vowel Hollering in a
Mob of Consonants* by James Bertolino, and *Crunch* by Lillian Ne-
cakov. I realized that the chapbook is the mouthpiece of feral
poetry. Do it. Just do it!

In her recent book of essays, *Animal,* Lasky again talks
about "the wild" in poetry: "a poem is wild when it is not
predetermined... like a wild thing that can breed endlessly, it
expands without asking" (54). Yes, a feral poet led by a wild
muse. A wolf, most likely. The astonishment of encountering a
timber wolf in the wild. This poem barely begins to express the
trepidation and wonder of that experience. It paces around it,
furtively, like a wolf coming across a human in the bush. And
then, finally, meeting her gaze.

Lupus Lamentis

This could be a poem about lupines, flowers for those rare occasions when the bite was not tenacious, the jaw didn't lock and the wolf sauntered on back into the woods.

Wolves are wary beasts anyway It was probably a black dog you saw running into the stream of unconsciousness that churns on and on after the electronic lock zaps shut.

Strepitus! clanging bars all down the hallway where the fluorescent lights are never, ever off.

This could be a poem about something else. *Tenebrae, crucifixio,* initiation rituals.

The young woman outside the church with the quiet voice and a knife.

Dumpster diver lamentis no ditched purses or body parts. Identities sawed away.

But it's not.

That late August afternoon walking alone in the parkland forest and suddenly the feeling of being watched. Across the clearing, forty paces and his head my shoulder height, thick tawny fur, wild amber eyes. *Timberwolves are reported to mate for life.* I remember writing and it seems like forever locked in his gaze when he bounds away, the black tip on his silvery tail a punctuation. An ending. A sign that I should not follow even though I want to. (Rowley, *Unus Mundus* 70)

Ecopoetics as Enactivist Poetics

At the Poetics Ecologies Conference in Brussels in 2008, I discussed poetry as an organic emergent system, based on Umberto Maturana and Francisco Varela's theory of enactivism – a cognitive neuroscience and systems biology model whereby organisms, species, communities and ecosystems co-emerge and evolve in a dynamic bio-tango of "reciprocal structural coupling." I was seduced by that poetic and sensual phrase, and enthralled by the concept of "autopoeisis," at once the emergence and continuance of life. It spoke to me of creativity and poetic emergence as well, and questions that I have often pondered: How do the poet's environment, mood, thoughts, interests, memory, history, sensory awareness contribute to or shape a poem? How does a poem create meaning? Is the connection of ecopoetics to enactivism merely metaphorical?

Importantly, in the enactivist view, language is not synonymous with communication, since it is experienced in tandem with emotions. Yet a barrage of indignant tweets can do nothing to quell the maelstrom of climate change and a polis that is becoming inured, no, acclimatized, to the escalating scale and incidence of raging floods, fires, drought, apathy, and mayhem. What, then, is the role of the poet in this era of climate devastation, techno-determinism, fake truth, commoditization of polis and politics? Sentinel? Canary in the coal mine? Activist? If so, how do we make our voices heard above the din of disinformation, the dystopian leaning backward and to the right? Perhaps by leaning farther back, reimagin-

ing a vestige of Platonic ethics, resurrecting *theurgy,* poetry as
ritual contemplation toward an embodied eco-poetic ethics.

At the Writing North symposium held at the University of
Saskatchewan in January 2020, Canadian poet and philosopher Tim Lilburn spoke about theurgy in the context of the
"poetic system" of the long long poem, where the poem itself
becomes an "ascending spiral of association and transformation." Intersecting cosmologies inspired by multiple muses let
free to run wild, "spread out, take as much space as needed"
to address the "sticky conundrum" of the poem. And he spoke
about how this type/form of poetry is formally tolerant, has the
capacity for orality, and comes alive in embodied performance
in a reawakening of Neoplatonist ritual. In resuming the quest
for truth, unity, soul in this technologically determined, politically and ecologically fractured present, it is poetry that can be
unabashedly political. Hence this poem:

SUPER TUESDAY

    Ultimately, the crows will call it –
        democracy
    a corvid call – Demaw Caw Craw See!

    Up here, above the border, no walls needed.
    Such a polite nation, if only
    we could reconcile
    get the railways rolling
        influence anything.
    Splurge and purge –

ripple effects as markets plunge, traders howl
tornados raze polling stations
   babies freeze at borders

of hell. Bots churn algorithms, turn voters
away from truth toward swagger
as old-white-rivals brandish rosebuds, quixotic
promises of health and normalcy for all.

As we all turn our heads,
don masks, stock up, quarantine, run –
   from Covid's mocking caw.

Living in Languaging and Emotioning

As a poet, my current scholarly interest in empathy, intentionality, and new media is rooted in my interest in language
– how it defines us as a species, constructs us as social and
moral beings, and how we co-evolve with changes in language that occur as a result of any new technology, whether
it be stick in sand, stone on tablet, ink on parchment, or hand
on keyboard. This is not evolution in the Darwinian sense, but
rather the rapid neurological and behavioural adaptation of
organisms (including humans) to their environments.

My interest in how poems evoke response in the reader,
listener, viewer then leads to the question, How is technology mediating and changing creative process, content, and
response? According to Maturana and Varela, we experience

the world through "languaging and emotioning," which not only modifies behaviour but also triggers structural change in the nervous system. "The phenomenon of communication depends on not what is transmitted, but on what happens to the person who receives it. And this is very different from 'transmitting information'" (1998, 196). This statement has provided a lush understory for my poetic and academic work, and I believe that it is important to feral poetics because it addresses the embodied nature of language and communication – even when digitally mediated. We are social animals, and like all animals, we communicate through myriad bodily cues, not just vocalization. Changes in "languaging" can trigger change in "emotioning" and resultant structural changes in the nervous system. Consider the despair and anxiety of a cyber-bullied teen.

Studies on the effects of social media on young people have been mixed and involve a host of external and psychological factors (Casas et al.; Chen et al.). Other studies have focused on the positive cognitive effects of online gaming (Shute et al.), while still others acknowledge the negative impacts of internet and gaming addiction (Bavelier et al.; Yuan et al.). Surprisingly few studies have addressed the issue of intentionality, embodiment, and social media. In existential phenomenology, embodiment refers to how we know and experience the world through bodily engagement and interaction. This begs the question, How is the body experienced in social media environments, and how does that affect intentionality, empathy, and our experience of self, other, and the world around us?

As Canadian scholar Max van Manen states, "Phenomenology, not unlike poetry, is a poetizing project; it tries an incantive, evocative speaking" (13). Can an understanding of embodiment and intentionality enhance response and engender
change? As a feral poet, I hope and believe that it can. Feral
poetics demands the incantive, enactive, evocative punch in
the gut, bite in the leg, lick of the face, howl, whine, purr, if, as
poets, we want our work to resonate, motivate, affect change.

Philosopher of technology Don Ihde discusses the phenomenology of imagination with respect to the arts, noting
the work of Heidegger on poetry, Sartre on literature, and
Merleau-Ponty on visual art. Importantly, "this phenomenology of the imagination shows, in contrast to many previous
theories, that the role of imagination is an irreducible function
of intentionality." For Ihde, imaginative autonomy consists in:
"its strict independence from other mental acts, from its sur-
roundings, and from all pressing human concerns... [and] the
freedom of mind of which imagination is uniquely capable"
(*Experimental Phenomenology* 108-9).

Recent neuroscience research confirms what poets and
artists have known all along: embodied imaginative autonomy involves a form of mind-wandering, or spontaneous
thought involved in creative thinking, which involves both
top-down "intention" and bottom-up "intentionality."

> The creative process tends to alternate between the genera-
> tion of new ideas, which would be highly spontaneous, and
> the critical evaluation of these ideas, which could be as con-
> strained as goal-directed thought in terms of deliberate con-
> straints and is likely to be associated with a higher degree of

automatic constraints than goal-directed thought because creative individuals frequently use their emotional and visceral reactions (colloquially often referred to as 'gut' reactions) while evaluating their own creative ideas. (Christoff et al. 720)

Yes. Feral poetics requires and demands imaginative autonomy: the mind must be allowed to wander. As Lasky states, "And while any kind of thinking makes the imagination embodied, it is the holy space of a poet's projected imagination, a space where new language can create new worlds, that does so most poignantly" (*Animal* 15).

OUR SULLEN ART

(excerpt)

the language of poetry has something to do

with the open mouth the tongue that jumps
up and down like a child on a roof calling
ha ha and who's the dirty rascal now?
the same boy sent to his room for punishment
leans from his window listening for animals
far away in the woods strains his ears to catch
even the slightest sound of rage but nothing howls
even the hoot of owls in the dusk is gentle
(Szumigalski 13)

Techne, Poiesis, and Being-in-the-World

When I was first reading Heidegger's *The Question Concerning Technology*, I was struck by the discussion of etymology; reading this passage, I experienced a eureka moment, like when I first understood verb declensions in grade school or algebra in high school. The word technology derives from the Greek *technē*, which, as Heidegger explains, "belongs to bringing-forth, to *poiēsis*; it is something poetic... *technē* is linked with the word *epistēmē*. Both words are names for knowing in the widest sense" (13).

"Technologies-R-Us" is my Heideggerian aphorism for Heidegger's concepts of *dasein, techne,* and *enframing.* Heidegger's philosophy of science has implications as to how we, as poets, continue to keep poetry relevant. As he states again and again, "The essence of modern technology lies in Enframing" (25). This concept of *Ge-Stell,* or enframing, posits that technology is not some "thing" or "other," but how we human beings understand, perceive, and experience the world through actions of challenging, producing, presenting, ordering, and revealing. But this essence of humans' relationship to technology is not just in "revealing," but in ordering nature into "standing reserve" for our use and disposal. (The image of oceans full of plastic waste comes to mind.) Heidegger's "danger" is that we become "enframed" by the very technology we create. Consider our dependence on electricity, Wi-Fi, cars. (On the prairies, we must travel vast distances to get from one town to another, particularly since the current conservative government has dismantled the publicly funded provincial bus system. I feel a feral growl emerging.) Consider how we are "enframed" by the commoditization and politicization of social media. And, as eco-poets, consider how enframing changes the relationship of *technē* to *poiēsis*. As eco-poets in the twenty-first century, we are technologically destined human beings. As such, our job is double-edged. How do we embrace technology as a means of poetic "enactivism" in order to "reprogram" commoditized intentionality?

Phenomenology of Technics and Programmed Intentionality

In *Technology and the Lifeworld,* Don Ihde describes technics as "the symbiosis of artifact and user within a human action" (72) and explains four existential relations in which humans interact with or engage in the world through technology. The first, *embodiment relations,* describes technology as an extension of the body or the senses, such as eyeglasses, a prosthetic device, a cane, a car, a joystick. Once the enhanced activity (walking, seeing, driving, gaming) is mastered, the device becomes transparent to the user and "withdraws"—we no longer notice the rims of our glasses or have to search for the right gear. Ihde's second phenomenological relation with technology, *hermeneutic relations,* refers to any action that involves interpretation of language, codes, or reading, where writing is not only technologically mediated language, but also an *embodied hermeneutic technics* (81-84). In his third category, Ihde describes the quasi-otherness of *alterity relations* in which "humans relate to technologies as relations *to* or with technologies, or to technology-as-other" (98). For Ihde, technological otherness is a *"quasi-otherness,* stronger than mere objectness but weaker than the otherness found within the animal kingdom or human one" (100). He describes the quasi-otherness of the video game as "the technology that fascinates and challenges" (101). Ihde depicts a fourth phenomenological engagement as *background relations,* a kind of "present absence" that functions in the background (108). In Saskatchewan winters, a furnace is necessary for survival; it is a technology that functions in the background, unnoticed until it malfunctions.

With respect to our use of social media, it is not the physical interface that is running in the background, fascinating us with its quasi-otherness—but the program running the application. According to Lev Manovich, the very programmability of new media makes it unique from previous technologies (47). When we interact on social media—with our real selves, avatars, or intelligent agents—not only are our actions mediated by embodiment, hermeneutic, alterity, and background relations, but *our interpretation and communication are mediated by the program we are using.*

21

mood portal
tongue in groove
swerve control
the better to steer you by
my dear

rough terrain, hurricanes
swelling in the brain
encephalitic ensilage
fodder for future
fits and storms

another kind of weather mapping

serotonin gauge
hormonal readouts
neuronal firing
electrocardio impulses
fluid secretion rate
blood flow

serous intentions
heinous intervention
(Rowley, *Suicide Psalms* 34)

Programmability defines how we communicate in online environments and how we interact in technologically mediated ones. Facebook's ever-changing privacy settings are one example. And when our attention becomes focused on the interface, rather than the use value of technology-as-tool, our involvement with technology, or "intentionality," changes. Swerve Control! Just replace "wristwatch" with "cellphone" in Ihde's prescient statement: "It may make little immediate difference if a wristwatch is worn as a fashion object, but if it successfully carries in its wake the transformation of a whole society into a clock-watching society, with its attendant social time, then a larger issue is involved" (*Technology* 129).

As Bernard Stiegler notes, it is not merely intentionality that is being programmed. Our attention—the core of collective individuation—not only is running at a deficit in our "network society of planetary proportions," but it also is being coded and commoditized into metadata (5). Stiegler refers to Katherine Hayles's differentiation between deep attention and hyper attention. Deep attention required in close reading and analysis – the mode on which humanities scholarship is based – is being replaced by hyper attention, "characterized by switching rapidly among different tasks, preferring multiple information streams, seeking a high level of stimulation, and having a low tolerance for boredom" (Hayles 187). Where is the time for reflection and imaginative autonomy?

Within the dynamic frameworks of intersubjectivity and enactivism, what happens when we become structurally coupled with(in) digital media with cyber-selves and cyber-others? We are co-evolving in and with technology, but biological evolution, unlike Moore's law, is not exponential. Neuroplasticity, or the ability of the brain and nervous system to adapt to rapidly changing environments, may outpace our ability to adapt physically, socially, and culturally. And what happens when the body is absent and our reading of the other is mediated by technics and modulated by programmed intentionality? How is empathy, along with other moral, social, and creative processes and decisions that require reflection, altered by immersion in social media environments?

FLOTSAM AND JETSAM

Quantum dot embedded in photonic crystal,
new ways of talking about what can't be seen
or heard. Listen carefully. Today's maps
of bodily processes genetic fodder for information
processing. DNA computing just a lick away, a blink
away, dust in the eye. In a jiffy. Swift as a burglar
with a jimmy, unlocking codes for red white brown
skin, crooked teeth, curly hair, high blood sugar,
hearts heavy or clogged, breasts lumpy or full,
passion strong or subdued, killer instincts.
How such a system is fabricated. Photonic crystals
resonate wave packets transmitted on bandwidths
auctioned on E-Bay. Important to regulate collusion
to increase traffic. Produce or sell flotsam and jetsam.
(Rowley, *Unus Mundus* 29)

Eco-Poetry and Biosemiotics

From enactivism emerged biosemiotics, where the sign,
rather than the molecule, is considered the basic constitu-
ent in the organization of living beings: "The chain of events,
which sets life apart from non-life, i.e. the unending chain of
responses to selected differences, thus needs at least two
codes: one code for action, (behaviour) and one code for
memory – the first of these codes necessarily must be analog,
and the second very probably must be digital" (Hoffmeyer
35). The genesis of both poem and biological process involves

multiple pathways and signals, which involve elements both of chance and of choice – the analog code of action and the digital code of memory.

The next step in my random walk took me to New York, and the eleventh annual Biosemiotics gathering at the Rockefeller University for Biomedical Research. There I met Marcello Barbieri, one of the pioneers of the discipline, as well as Donald Favareau, who has applied Peircean semiotics to social neuroscience and the mirror neuron system. I like to describe the concept of *epigenesis,* a basic tenet of enactivism and biosemiotics, as a dance:

> The self-assembly of polypeptide chains into three-dimensional proteins is only the first step in the intricate choreography of cells that combine to form molecules, which combine to form organisms. Although genetic information is the first step, the complex tango involves interpretation, timing, and an ongoing increase in complexity – or *epigenesis.* (Rowley, "Ecopoetics")

What is the nature of poetic emergence – the epigenesis of a poem? "How does the poet read and interact with her environment, or semiosphere, in order to translate emotions, memories, sounds, smells, disconnected images into the phonemes, syllables, words, lines, and stanzas of a poem that resonates with a reader/listener? By what mechanisms does a poem evoke emotional or physiological response?" (Rowley, "Mini Manifesto"). Do all poems have these mechanisms? Is poetic code organic? What does nature (biology) have to do with this?

BIRD SONG JET STREAM

Glimpses of morning. Glimpseglimpseglimpse, nut-
hatch noise. Perfect bird language –
    such awkward English. Crows call overhead.
Awkawkawkawawwwwk.

    Birds bicker at the feeder. Fighter jets etch the sky
with lace. No –
    beware of used-up metaphor, overextended lace.
More like

    Egyptian blue embossed with white, except more
copper oxide in the blue,
        more smoke in the white.

    Forest fires rage, evacuees run, smoke smudges out
the sun.

    I was going to write about the mystique of flight.
Take-offs and landings.
        Soaring above the clouds. The push of wind under
wings. Thrust and drag.

    Old photo and bones. Amelia Earhart's laughter, dis-
tress calls caught in the tail wind.

    Back at the feeder, chickadees and nuthatches land
and lift-off. Birdseed
    a bumper crop. The cat waits by the fountain for the
birds to drink. (Rowley, *NumenRology* forthcoming)

Eco/feral poets are not surprised by the concepts behind biosemiotics. Of course, molecules, organisms, and animals communicate in and with their environment. We hear them. We are constantly sniffing out signs. Epigenesis is a convergent increase in complexity, whereby a system – whether cell, organism, or language – is created from incomplete information. The germ, or zygote, of a poem can be a single word, image, or memory, and often a poem emerges unbidden, unfolding its own strands of meaning. Feral poetry seeks this kind of epigenesis in poems that sidestep intent, misbehave, mutate syntax and grammar, mangle metaphor, mash content, and leave the poet and reader charged with a quantum leap of energy experienced viscerally, in the body, at a cellular level.

In a wonderful collection of essays titled *Material Ecocritism,* Wendy Wheeler acknowledges the epigenetic nature of creativity: "The experienced scientist is grasped by a significant detail, without yet understanding its full implications, just as the experienced reader is similarly grasped by a symbol of some kind that equally seems significant. Both are on the lookout for emergent patterns" (72).

Eco-poets need to put the reader/listener "on the lookout" for these emergent patterns. Feral poetics, and most creative and scientific discovery, involves synchronicity. Elements of chance and choice. Speaking of synchronicity – whom should I meet at the biosemiotics conference but Stephanie Strickland (who lives almost across the street from Rockefeller U). I was thrilled, as I greatly admire her work both on the page and "through" the interface – for her astounding breadth of knowledge, depth of insight, and innovation of language and digital poetics. Also, for what I see now to be a mastery of cerebral-visceral, neuro-phenomenological coupling. The Feral.

[EXCERPT FROM L'UNA LOSES]

0.4

of gambling.
L'una
loses. Luminous, lingering, dropping

0.3

her net, her cut
spool of star-
stained ocean, Channel foam, the Roman

0.2

sea, cliffs of Manhattan.
I will be added and united.
Failing, falling

0.1

behind,
below, inside, the crashing, faintly
smoking surf, she

0.0

fades

to day, to dawn, to
gone, to Dis-

    0

appeared. (Strickland 49-50)

As Wendy Wheeler states, "Not only does [biosemiotics] put humans and human *poiesis* and *techne* back in nature where they belong as evolutionary developments, but it also erases the false sharp modern distinction between mind and body, nature and nurture, and materialism and idealism" (69).

Neurophenomenology, Embodied Simulation, and Being-in-the-World

The discovery of mirror neurons in 1996 by Vittorio Gallese and Giacomo Rizzolatti and their colleagues, and subsequent research on the mirror neuron system (MNS), was groundbreaking not merely for furthering understanding of social cognition, but also for reinvigorating phenomenological enquiry and reuniting body and mind, or "being-in-the-world" with "being-in-our-heads." In "Mirror Neurons and the Social Neuroscience of Language," Gallese notes that the MNS has been used to explain imitation, mind reading, our understanding of action and intention, and empathy. He has also investigated the involvement of the MNS in aesthetic experience and language.

Gallese and colleagues discuss empathy in light of embodied simulation, a phenomenological, physiological, and neurological understanding of another's emotions, actions, and intentions, which leads to "intentional attunement." In contrast to the top-down, cognitive theory of mind, embodied simulation is a bottom-up, "mandatory, nonconscious, and prereflexive mechanism" and "a prior functional mechanism of our brain" (Gallese et al. 143).

> Gallese's concept of embodied simulation... is a neuro-phenomenological approach to social cognition where one's bodily self resonates with other selves in an enactive, intersubjective, pre-reflexive, multimodal sensorial dance of understanding. There is something innately feral about this—the body reading/responding to body. (Rowley, "Toying" 11)

In feral poetics, whether one uses cut-ups, programming code, video, mathematical formulae, or hypertext, the content and media intuitively and intentionally influence how the work is experienced by the reader, listener, viewer. As Hayles states in referring to Hansen, "the role of embodied perceiver (is) not only a necessary site for reception of digital art work(s)... (they) literally do not make sense without taking embodiment into account" (*Electronic* 36).

In recent research, Gallese has collaborated with artists, curators, and dancers. He discusses how the notion of Einfühlung, posited by German philosopher Robert Vischer in 1873, has been translated as "feeling in, or feeling into" and says, "This account of art perception implies an empathic involvement, which, in turn, encompasses a series of bodily reactions of the beholder" (Gallese and Di Dio 2012). The discussion of art, empathy, and the embodied perceiver has moved from the domain of critical theory, digital humanities, and computer science to the realm of social neuroscience and neuroethics.

Some Posthuman Philosophical Musings

But what about ubiquitous computing and other techno-logical toys/tools being developed to program our intention-ality? In the rush toward the posthuman, eco-poets need to be asking the question raised by Canadian physicist and philos-opher of science of Ursula Franklin: "[For] whose benefit and at whose cost?" (124). How can the knowledge of embodied simulation and neurophenomenology be used to reprogram commoditized intentionality in posthuman environments in order to evoke a visceral, emotional, embodied call and re-sponse, where *techne* and *poeisis* (technology and poetry) are merged to created *episteme,* or knowing in the widest sense? How can ecopoetry recapture a sense of wonder?

Philosopher of science, gamer, and proponent of ob-ject-oriented-ontology Ian Bogost posits that the way to redis-cover wonder in a posthuman world is through "thingness," where human, the potato, rock and beast share an equal claim to subjectivity. "Things are independent from their constitu-ent parts while remaining dependent on them" (23). An echo of enactivism and epigenesis can be heard in this statement. And in his essay "The Liminal Space between Things," Timo-thy Morton discusses the "thingness" of art, including poetry and talks about the nonspecifiable "thereness" and "disturbing spontaneity of epiphany," in both the creative process and the "ah ha" experiencing of it. He describes the "epiphany of beginning" as a "crack in the real," where "something has already happened." Epiphany is central to feral poetics, where "all kinds of nonhumans are already involved in the existence of a poem" (271).

I believe ecopoetry can recapture a sense of wonder by evoking/invoking the feral howl.

So What Is Feral Poetics Anyway?

- Feral poetics is enactivist poetics—brought into being by the poet in tandem with her environment, body, mind, psyche, knowledge, craft, and polis.
- Feral poetics demands awareness of both "mindbody" and "machinebody."
- Feral poetics is a neuro-phenomenological creation and response engaging both bottom-up visceral (embodied, pre-reflexive, intuitive) meaning/making with top-down cerebral (reflective, intentional) understanding.
- Feral poetics infiltrates the viral stream of new media. Surfs it. Feeds off of it. Plays with it but sets the rules of the game.
- Feral poetics cares about what happens to the person who "receives" it.

- Feral poetics embraces hybridity and has a healthy respect for the "other."
- Feral poetics turns heads and changes minds—modifying thought and setting-to-action changes in behaviour in-the-(real/physical) world.
- Feral poetics is the antidote to programmed intentionality.

The feral, embodied eco-enactivist poet weaves words into
code as a subliminal digital understory, soft underfoot, nour-
ishing, rejuvenating, yet mutable and adaptable. We raise our
collective, networked, feral, viral voices to be heard and felt in
order to create a TechéPolis in which we and other species can
survive and thrive.

In closing, the best feral poets are disobedient in the best
sense. I think of Alice Notley, Anne Szumigalski, and Saskatoon
poet and Griffin Poetry Prize winner Sylvia Legris, who is a very
disobedient rebel when it comes to prairie realism. I want to
end on this poem, haunted as it is with the notion of the body
as alien thing in a disturbed collective psyche and ecosystem.
If Anne Szumigalski's feral muses were angels, Sylvia's are
birds, with their beaks and tongues and loud shrill voices.

AGITATED SKY ETIOLOGY

STUMPED SKY (QUESTIONS OF MISSING WEATHER
AND BIRDS)

2

Lack. Lack. Lack. Lonely ducks plead for rain but rain
rain's gone away and the trees
have pulled inside themselves (multi-stumped
and trunks
a frazzle of missing leaves).

Weather is numb. Nonsensical. The sky all thumbs
and fingers falling.
What's the point here, what's the point there:
unceasing questions.

Clouds a flummox of fluster. Flux. Ice miasma.
(Second nature
a temperate climate preceding storm.)
(Legris 71)

Works Cited

Banting, Pamela. "Making Scents: Signature, Text, Habitat." Poetics Ecologies: Nature as Text and Text as Nature Conference, May 2008, University of Brussels, Belgium. Keynote Address.

Bavelier, Daphne, et al. "Brains on Video Games." *Nature Reviews Neuroscience*, vol. 12, no. 12, 2011, pp. 763-68.

Bogost, Ian. *Alien Phenomenology, or What It's Like to Be a Thing.* U of Minnesota P, 2012.

Casas, José A., et al. "Bullying and Cyberbullying: Convergent and Divergent Predictor Variables." *Computers in Human Behavior*, vol. 29, no. 3, 2013, pp. 580-87.

Chen, Liang, et al. "A Meta-Analysis of Factors Predicting Cyberbullying Perpetration and Victimization: From the Social Cognitive and Media Effects Approach." *New Media and Society*, vol. 19, no. 8, 2017, pp. 1194-1213.

Christoff, Kalina, et al. "Mind-Wandering as Spontaneous Thought: A Dynamic Framework." *Nature Reviews Neuroscience*, vol. 17, no. 11, 2016, pp. 718-31.

Clark, Hilary. "Feral Muse, Angelic Muse: The Poetry of Anne Szumigalski." *The Literary History of Saskatchewan*, edited by David Carpenter, vol. 1, Coteau, 2013, pp. 193-202.

Franklin, Ursula. *The Real World of Technology.* CBC Enterprises, 1990.

Gallese, Vittorio. "Mirror Neurons and the Social Nature of Language: The Neural Exploitation Hypothesis." *Social Neuroscience*, vol. 3, no. 3-4, 2008, pp. 317-33.

Gallese, Vittorio, and Cinzia Di Dio. "Neuroesthetics: The Body in Esthetic Experience." *Encyclopedia of Human Behavior.* 2nd ed., Elsevier, 2012, pp. 687-93.

Gallese, Vittorio, et al. "Intentional Attunement: Mirror Neurons and the Neural Underpinnings of Interpersonal Relations." *Journal of* the American Psychoanalytic Association, vol. 55, no. 1, 2007, pp. 131-75.

Hayles, N. Katherine. "Hyper and Deep Attention: The Generational Divide in Cognitive Modes." *Profession*, 2007, pp. 187-99.

Heidegger, Martin. *The Question Concerning Technology and Other Essays*. Translated by William Lovitt, Harper & Row, 1977.

Hoffmeyer, Jesper, and Claus Emmeche. "Code-Duality and the Semiotics of Nature." *Biosemiotics: Information, Codes and Signs in Living Systems*, edited by Marcello Barbieri, Nova Science, 2007, pp. 27-64.

Ihde, Don. *Experimental Phenomenology: Multistabilities*. 2nd ed., SUNY P, 2012.

⸺. *Technology and the Lifeworld: From Garden to Earth*. Indiana UP, 1990.

Lasky, Dorothea. *Animal*. Wave, 2019.

⸺. *Poetry Is Not a Project*. Ugly Duckling, 2010.

Legris, Sylvia. *Nerve Squall*. Coach House, 2005.

Lilburn, Tim. "The Long Long Poem." Turning West: Writing in Western Canada, Writing North Conference, January 24-25, 2020, St. Andrews College, University of Saskatchewan. Keynote Address.

Manovich, Lev. *The Language of New Media*. MIT P, 2002.

Maturana, Umberto, and Francisco Varela. *The Tree of Knowledge: The Biological Roots of Human Understanding*. Shambhala, 1998.

Morton, Timothy. "The Liminal Space between Things: Epiphany and the Physical." *Material Ecocriticism*, edited by Serenella Iovino and Serpil Oppermann, Indiana UP, 2014, pp. 269-81.

Notley, Alice. "The Poetics of Disobedience." *Poetry Foundation*, 15 Feb. 2010, https://www.poetryfoundation.org/articles/69479/the-poetics-of-disobedience.

Rowley, Mari-Lou. "Ecopoetics as Enactivist Poetics." *Poetic Ecologies: Nature as Text and Text as Nature*, edited by Franca Bellarsi, P.I.E.-Peter Lang, forthcoming.

Rowley, Mari-Lou. "*Unus Mundus:* Mini Manifesto." Academia.edu, 3 April 2013.

    https://www.academia.edu/3287507/Unus_Mundus

————. "Toying with Intention: Embodiment, Empathy and Programmed Intentionality in New Media." *The (Un)Certain Future of Empathy in Posthumanism, Cyberculture and Science Fiction,* edited by Elsa Bouet, Brill, 2015, pp. 1-16.

————. *Numenology,* Anvil, forthcoming.

————. *Suicide Psalms.* Anvil, 2008.

————. *Unus Mundus.* Anvil, 2013.

Shute, Valerie J., et al. "The Power of Play: The Effects of Portal 2 and Lumosity on Cognitive and Noncognitive Skills." *Computers and Education,* vol. 80, 2015, pp. 58-67.

Stiegler, Bernard. "Relational Ecology and the Digital Pharmakon." *Culture Machine,* vol. 13, 2012. svr91.edns1.com.

Strickland, Stephanie. *WaveSon.nets/Losing L'una.* Penguin, 2002.

Szumigalski, Anne. *A Peeled Wand: Selected Poems of Anne Szumigalski,* edited by Mark Abley. Signature, 2010.

van Manen, Max. *Researching Lived Experience: Human Science for an Action Sensitive Pedagogy.* SUNY P, 1990.

Wheeler, Wendy. "Natural Play, Natural Metaphor and Natural Stories: Biosemiotic Realism." *Material Ecocriticism,* edited by Serenella Iovino and Serpil Oppermann, Indiana UP, 2014, pp. 67-79.

Yuan, Kai et al. "Microstructure Abnormalities in Adolescents with Internet Addiction Disorder." *PLoS ONE,* vol. 6, no. 6, 2011.

Fasciae, Fauna, Faux Pas: Capital Beyond the Noise

ARPINE KONYALIAN GRENIER

> to be lost to numbers
>
> to not let greed define emotion
> glug merely passage
> compassion
>> willed option
> devotion

I am interested in change. I'm interested in the will to align invention with transformation, a will ready to face what follows – the risk, the reward, the unknown. Such will would register what matters like some Higgs emanating mass, and we, its beneficiaries resisting norms to redistribute and recalibrate as necessary. To resist is to create, changing the state around a state, revitalizing outlook, capacity.

The outlook on capitalism is grim (unless we rearrange (perturb?) the conceptual grids that define capital). Fear, the prime component of our survival mechanism, has gone mad;

disconnected from the natural laws of the universe, it has been nurturing greed, consequently drastic and irreversible havoc. Extract nothing from history but all the whats and hows – pink silly agglutinations decidedly colliding. Our very large collider at CERN (the European Organization for Nuclear Research) vouches for that, a tear at the core or the edge, varieties of symmetry as the symmetry within variety, color charge, electrical charge, magnetic form factor against momentum. Autonomy, complexity and reward hand in hand we'll measure, report, select and sail as we go.

About when we turned sentient and sensible, one hundred thousand sages for change intervened to convene. Evolutionary pathways host breaks, cracks and ruptures in a seemingly harmonious world, they said. Theorizing is as futile as rationalizing because selection pressures are inconsistent and often do not make sense. Consider "self" part of "the other," what may be noise will subside. All seems fair then, the purgative, the illuminative unitive. Do not look at the whole to annotate, look at parts and pretend. Think of how matter behaves differently under different conditions. You are such matter. Center and periphery are mythical allusions. Speech exerts to overcome excess. Mind angle and trajectory as switch and turn are about to follow. As love does not fail and sentences do not restrict the soul to a mere parsing of words.

is this a revelation?

maybe

To consider ecologies of change plastic. Levitation surmis-
es. Phylogeny is molecular but also neurological, inimitable
but additive. There are species that molt, and those that do
not. Snakes were four legged once. Mice and flies carry the
same linearity of genes in the fore, mid and hind sections of
the brain. Narrative interrupting narrative unfolds mornings
at Ur, fractures, ameliorates itself. Segmentation in annelids
and arthropods evolved independently but also in convergent
manner, not corresponding but convergent (versus homologous, which harbors the genealogic). Containment tactics
distract the masses while will that has pulse ensures access,
trade, capital.

> I will not fit into or figure or flee from the human
> I will feel It Instead

> the more taxa are related the less new species arise

> when we limit a field it deepens says DeJong
> a spandrel is not perceptually constant

I imagine you a raptor at the door, watching a rat ponder
thoughts it had a long time ago, body mapped unto genome
in multiple dimensions, gene to chromosome to organ, a uni-
fied system whereby division of labor replaces the aaaahhh
of being born, dendrites and axons part of a dynamic similar
to that of weights and pulleys. There are of course sensory
and motor connection costs, innate configurations based on
physical and mathematical orders of the universe. There are

also compromises, improvisation, the ad hoc. Glomerulus to gene receptor to protein to space recognition and territorial behavior, the over six million years old neuro-modulators in the brain provide to the suchness of memory, reality, phenomenology and experience, seeking the poetics behind clarity, charity, humility, humidity, forgiveness and compassion. Are we about a living thing or the human?

> what about magic?

> it is a small and round world indeed

> love wills through still

> long time bristled edges effort/proclaim/abandon

> suchness flashing by is it?

Constantinople to Istanbul, Dicranagert to Diyarbekir, we'll revisit nativism because the inchoate is in our backyard. Will optimization safeguard our lot in the cosmos? Fat chance outhouse institutions will break into or up the pristine setups we have fashioned for ourselves. We are facing the trans-national, possibilities lurking. Life is a possibility, space is opportunity. There's more space when gravity is at work, silently. Silence narrates as narrowness imposes on depth. The I feels intent, the rose extends.

> there has been a cut therefore dialogue
>   tightness to lightness reveals

– is it true? (authentic)
– is it timely? (useful)

– does it have roots in discernment and goodwill?
– am I hearing and speaking effectively?

Modulate parametrically, sort and figure direction as you go. Experience alters perceptual space, homologous matter, self or otherly architected matter, inwardly or outwardly generated fractures in and out, congruent or incongruent. Angular distributions and velocities have bearing, the orthogonal too. Change follows connection, and when powers (agendas) are in optimal orientation to one another, newer topologies and innovative notions emerge. (As in, we are told light travels in straight lines, yet under certain conditions (solutions) light is observed in knots and links.)

Difference is homeless. Otherwise, possibilities are pigeon-holed drama, and words like coercion, disease, life, love and power are politicized. How we are captives then, the anterior temporal lobe of our brains rendered useless, drowning in cerebrospinal fluid.

Vox populi thunders to remind and foretell, to sing to foretell. *Cantare.* Go for it, categorize, generalize, learn from nature. Flowers are there for the bee, yet sucking nectar without pollinating the flower goes against nature. Perception to reception is redemption capital, unlike perception to production. Mindful modeling would inspire post-colonial ecologies. Our innate capacity to create egalitarian relationships is enormous because we all come from the humiliatable – a materiality that is luminous and forgiving of matter, more like a developing act that creates patrons and heirs on both sides of the equation. A collective intelligence is our only chance with humanity. The highly touted concepts of freedom and democracy would be considered mutually exclusive if it were not for relationality which gives them shape, possibility. Each impact an entrance, sudden departure, vigilant eye.

Entanglements are necessary, articulation between fragments, spaces and times, approximations and optimizations too. Do not be alarmed, eviscerate, they say. The forge of language discloses the truth of our feelings, the sensory knowledge of our affects. There are no protagonists, antagonists or narrators where it lands, only practitioners of an eco-poetics after inclusion. The price of a miss is what the how(l) dictates. Fixation and exploration are one and the same *hajar* to *hijrah* drift, stone to migration tremor, saccades. A high resolution view in a dim room with chin rests awaits. Fracture font, what surrounds it is what volume, sound, timber and feeling express. Serial and non-serial processes congregate. Attention alters the path of the process, colorful neutrinos crisscrossing/ penetrating the translucent and opaque ecologies, providing perspective.

Elohim contracted to allow for creation, say the one hundred thousand sages. Feed on mass gaps and symmetry breaks, it is a small and round world indeed, codignly yellow. Remember Ur. Limits are constantly being negotiated all the way to the cerebral cortex, frequency specific. Elegant form often responds to ugly challenges, which is when the system starts cooking, "at the edge of chaos," say Piattelli-Palmarini and Uriagereka in *The Evolution of the Language Virus* (373).

witness truth reality fact are states not acts
hypotheses unravel narratives

revel in *al muqarnas*

Ecologies of fracture are telling vistas empathy defines, illuminates and recreates. Blighted and blistering neutrinos of hope and desire festoon the burgeoning architecting, moral judgment and will reaching for newer modalities in the cerebral cortex where a world history of erasures is screaming, I will contract no longer except knowingly pernicious yet remarkably yielding to the irony of time. Migration is movement that refreshes, is maya, full of curiosity and hope. Evolution, its shadow. Said Leopardi, artifice lacking truth is valuable and effective but truth lacking artifice is impotent.

facsciae fauna *faux pas* greet me
feel me feed me

Tao, metaphor, origin. Narratives. Individuals do not evolve, societies (populations) do, and they do so with small changes, they declare. By chance or otherwise change occurs and is sorted over time. Neither food nor fame or fortune will provide access to the inner power we possess, the one that frees us from the slavery we have come to live by, the pervasive and addictive slavery that sedates the self under the guise of some external power we seem to think we must attain or give in to. Dare we amble by? As the ambivalence of intent (destination) binds process challenging connectivity at the purposeful and the productive, we come to realize cultural legacies matter, but they are not indelible.

we define as we go

narratives come up
narration continues

Language helps and works because it is never only about language, and because it ultimately fails us as it mirrors history, theory, memory and our very being. We'll work with language, we'll use it but not adhere to it. And if we allow ourselves a lesser lean on fundamentals, we'll experience it differently because when the word hits the page (or vocal chords), it has already created more (or less) than itself. Ur informs, there's hope and wishes for every reality gone mad, and reverence, not to respect but to honor the very histories that have brought us to where we are, here to honor the warmth of our blood, the cognitive and the normative, not part of a qualified culture nor speaking on behalf of one, busy with dimension while weary and leery of its shadow, calling, calling at an unbearable proximity to the need to do something about all that does not follow the routine of civilization, the functions of circumscribe – control – eliminate; calling to fuse what is outside of time with what is within, often at the expense of meaning or syntax, to reconcile with the self as if.

secondary syllables undermine love
duly splayed for you and me
virtue is overdue

                I love you

do you love me?

What questions do we ask then, and where and how do love and empathy come in as we embrace the potential and limitations of personal power. There is no hierarchy there, no pleasure in hierarchic power either. We, from the prehistoric to the agrarian to the industrial and the techno, language and empathy our fairly recent traits (as we've come to realize that we, after all, need one another).

We need a collective that seeks unfettered communication, exempt from historical, hierarchical and culturally exclusive expectations in order to negotiate the multivariate nature of life, love, reality – the human. Pat answers and simplified narratives are defenses we've been using to anesthetize ourselves against a life in synchrony with nature and nurture. They are out.

ode to the looming main tenant

the subtle causal cosmic
        politely insidious

refer/infer/confer/offer/defer/profer

Think model, not data, not numbers. Extract, group, relate and classify to be rid of noodling minds thinking, life's not what all it was made to be but, it is something, rid too of mainstays like hearts filled with fear (greed) cuddling music unto death; the pull, capital, for example. Why do I write you then? Oh but it is, how do I write you. No, why is first, the cello-spastic chain of de sire. De who?

> Doctor Who I love you Doctor What what's with gaps
> ever luminous Doctor Where proceeding dorsal or
> how to Doctor How Doctor Why must
> because Doctor When says now

>   corny isn't it?

> But then now is the time for all women and men to foster
> fracture as found territory one harvests

> lap lap the river bend brew
> list yourself as meanwhile
> truth is missing itself

> I hear you
> slovenly

An earth spot longing for a frame. *L'arbre et la glycine* (See Zins). Closure. I have no frames for you but the room is ready. Because I saw space flutter by, charmed but also charming. Anomie created anomaly, I am not indifferent to you. On the other hand, *Aghd.* I read you as I, dirt stained and twined of light write you, with capital. For so and lo, for Mama. Where is Mama?

Hear me spread for manuscripts, inside out or outside in barking for the space between death and the dying. I crouch by, sleep, push, taper. The document. It is breathing. There is no Mama. Silk, paper, gunpowder at the concession stand. That's you, a concession stand, one, Eve related conflagrates to, another bridges or insures, random phase. The poetry of life insists fractures are dwelling too. Go there, show up, local to globally focal. *Al muqarnas* rules, how well it redeems as it illuminates.

opt out and wed the op eds

Did we ever talk this way or is this another candles-plates-wine construct or candles-wine or just candles and a public throat beyond the pale suspect of brave horses – cool – rush – red – bullet red – where light is mere system among other corpuscular attitudes, mottled song pressed for shine in the husk of matter, the document. Word to phrase to sentence, paragraph, stanza, the physical world translates the biological, the social, the familial and you and me. Think of the job ahead, the floaters and sing-song for, a dimmer and dimmer you living your death parts retaining allude, allude. The abnegation genders. Capital. To appropriate because of the more better and the very many – matter missing out on light, knowingly ceasing to breathe. One wants to charm, confuse, display, reuse what's snared in-between. *Confiture.* The abnegator's. I, Euphrasiac Eve, debarking, unbarking "you see?," "see what

I'm sayin'?," "like, like...," "in your face," "in the belly of Baal," a newer space-time I will you in. Scream, drums, chainsaw into the river. Euphrates, Eurydice, bellied in.

Works Cited

Piattelli-Palmarini, Massimo and Juan Uriagereka. "The Immune Syntax: The Evolution of the Language Virus." *Variation and Universals in Biolinguistics*, 2004.

Zins, Céline. *L'arbre Et La Glycine: Poéme*. Gallimard, 1992.

Walter Benjamin, John Cage, and Kenneth Goldsmith: A Weather Report

PETER JAEGER

EXPOSÉ

The writings of John Cage play a central role in the citational, intertextual poetics of postwar writing and art. This essay focuses on Cage's *Lecture on the Weather* (1975), as well as on American poet Kenneth Goldsmith's *The Weather* (2005). The essay also takes a cue from these writers' innovative poetic practice, in order to explore the notion that critical writing can become, in the words of American poet and critic Julianna Spahr, "the subject of its own engagement, giving itself over to the dangers and fluidities and challenges of that possibility" (7-8). To this end, the essay employs such poetic tactics as modularity, disjunction, radical juxtaposition, intertextuality, and plagiarism—all devices used by the poets in question, albeit in very different ways. And so along with its discussion of Cage and Goldsmith, a secondary feature of this essay is an exploration of critical montage, based loosely on Walter Benjamin's citational method in *The Arcades Project.* I hope to open up an alternative mode for the production of knowledge by demonstrating the effectiveness of Benjamin's method in the field of contemporary poetic analysis. To paraphrase Benjamin, this form of critical work is aimed not at producing a knowledge of poetry, but at producing a knowledge that grows out of poetry.

BENJAMIN/NOTHING

"The method of this work is literary montage. I have nothing to say" (*Arcades;* qtd. in Rolleston 16).

CAGE/NOTHING

"I have nothing to say and I am saying it and that is poetry as I need it" (*Silence* 109).

GOLDSMITH/NOTHING

"I'm interested in a valueless practice. Nothing has less value than yesterday's news" ("Uncreativity" 1).

CENTENNIAL/ BICENTENNIAL CELEBRATIONS

In 2012, the year of John Cage's Centennial celebrations, I was fortunate enough to attend a performance of Cage's *Lecture on the Weather* at the Music Gallery in Toronto. This piece was originally commissioned by the Canadian Broadcasting Corporation in 1975, in order to celebrate the American Bicentennial. Cage responded to the CBC's invitation by composing a score for twelve "American men who have become Canadian citizens" (*Empty* 3) – a reference not only to America's historical past, but perhaps also to the fact that Canada, and especially Toronto, had provided a haven for many American draft dodgers during the Vietnam War.

MECHANISM

The score for *Lecture on the Weather* was devised by recycling Thoreau's *Walden, Essay on Civil Disobedience,* and *Journal* to *I Ching* chance operations. Cage began using chance operations in 1951, when he was given a pocket version of the *I Ching* by one of his students, the composer Christian Wolff. He used the text to determine such compositional factors as the number and duration of sounds in a musical composition, or the choice and placement of words in writing. It is important to note that Cage was "less interested in the *I Ching* as a book of wisdom than as a mechanism of chance operations that produces random numbers from 1 to 64" (Lewallen 235). The point of relying so heavily on the oracle was not to receive arcane solutions or esoteric knowledge, but to compose intricate scores based on numerical values, and to reconfigure subjective intention by limiting choice.

AN EXCERPT FROM LECTURE ON THE WEATHER: CAGE OR THOREAU?

"I lived alone, in the woods, a mile from any neighbour, in a house which I had built myself, on the shore of Walden Pond, in Concord, Massachusetts, and earned my living by the labour of my hands only. I lived there two years and two months" (Cage, *Lecture on the Weather;* Thoreau, *Walden* 3).

DMZ

In *Lecture on the Weather*, Cage paired his modified Thoreau texts with recordings of weather phenomena, such as wind, rain, and thunder, as well as with the projection of a film of flashing negatives of Thoreau's drawings, which represented lightning. Cage's choice of Thoreau is also significant; in 1967 Wendell Berry introduced him to Thoreau's *Journal*, and Cage writes that at that time he was "starved for Thoreau" (*Empty* 11) and that in reading Thoreau's *Journal* he discovered "any idea I've ever had worth its salt" (*Empty* 4). Cage produced several texts based on Thoreau's writing, including *Lecture on the Weather* and *Mureau* (1970), the title of which comes from condensing the words *music* and *Thoreau*. Thoreau in Cage's hands becomes a metonymy for natural processes, in which the treated text is a material thing, another element in the field of nature. The Cage Centennial performance in 2012 – as opposed to the 1976 Bicentennial performance – moved away from Cage's original call for American performers by including performers who were both American and Canadian, as well as male and female. This latter performance seems more in line with Cage's 1972 claim that syntax is the arrangement of the army, and that as we move away from syntax, "we demilitarize language" (M). Cage employed this metaphor for language during a period when America was still enmeshed in an increasingly unpopular war in Vietnam; the demilitarized zone between North and South Vietnam was an area that by treaty was not supposed to permit any military activity, although in reality it was bombed by the Americans, clustered with land mines, and used as a battleground at various points during the war. In theory a DMZ functions like a neutral territory, free from the administration and control on either side of its borders. To present the nonsyntactic use of language as a form of DMZ is to metaphorically compare linguistic freedom with social revolution: "We begin to actually live together, and the thought of separating doesn't enter our minds" (*Mureau*). *Lecture on the Weather's* 2012 performance, with its use of male and female American and Canadian performers, perfectly models the type of DMZ Cage called for in his writings about language itself. This shift in the national identity and gender of performers is an interpretation on the part of Paul Walde, the multimedia artist and composer behind the Toronto performance; although Walde maintained the rigorous, chance-based constraint of the original score, his decision to include a variety of performers is valuable, inasmuch as it resituates the text outside of the specific Bicentennial context of its original performance. We could say that Walde retains respect for Cage's compositional method, while simultaneously upholding his desire for social revolution, in which the thought of separating national and social borders "does not enter our minds."

Digital Archives

"No one expected the Vietnam War to play out as it did. With thousands of young men fighting to the death overseas, another group of American sons fled their homeland and journeyed north to Canada. As the battle raged on and the antiwar movement divided the United States, draft dodgers and deserters struggled to forge new lives for themselves. Seeking sanctuary and the opportunity to make a difference, they changed their adopted country unquestionably. (Note: Some clips contain explicit language.)" ("Seeking Sanctuary").

POLITICS

The uncanny, overlapping voices and climatic (climactic?) effects exhibited in Cage's *Lecture on the Weather* take on the quality of a dream; voices narrate Thoreau at different paces and volumes, and the specificity of each voice is lost amongst the excessive disjunction of the whole. Here we might turn again to Benjamin's notion of the dialectical image. On one hand, the dream-logic of *Lecture on the Weather* is rooted in the past, in the archaic language of Thoreau and his desire for ecological harmony and civil liberty. Simultaneously, the lecture is aimed toward a future, an egalitarian and anarchistic desire for social freedoms to come. In Michael Calderbank's account of Benjamin, "The dialectical image appears at a moment when future elements of past experience flash tantalisingly into the 'now-time,' retroactively disclosing unknown layers of significance which have been secreted in the object" (6). Cage's text uses the past (Thoreau) to unfurl the future—or at least, to unfurl a desire for the future as utopian collective, flagless, without the exclusive borders of the nation state: "I dedicate this work to the U.S.A., that it become just another part of the world, no more, no less" (*Empty* 5). In the dialectical image, "politics attains primacy over history" (*Arcades* 388-89).

DREAM

"The dream, for Benjamin, is an index of freedom—our social dreams indicate our social utopias.... Historical awakening is an aim of the project as a whole" (Leslie 2).

NATURAL PROCESS

Lecture on the Weather foregrounds the boundaries between the sociolinguistic construction of nature and the formal, material processes that are found ubiquitously in the natural environment. Rather than depicting the ecopoetical themes of decay, entropy, and transformation by pointing to these characteristics in an extratextual referent, Lecture on the Weather emphasizes ecological unpredictability and dynamism in material form. Cage employs a surplus of recycled information about nature, to the point where that information cancels out its own clear transmission. For Marjorie Perloff, *Lecture on the Weather* is mimetic of natural processes. "Whatever its venue," she writes, *Lecture on the Weather* is "essentially a mimetic text, one that simulates 'weather'" ("Moving" 76). Cage's text enacts the weather by moving through various climatic conditions, all of which are yoked to the recycled language of Thoreau. And just as weather moves across national boundaries and time zones, Walde's 2012 Toronto interpretation moved the Lecture away from its original situation as a specifically "American" performance in Canada, and toward a new, borderless zone that is performative of natural processes. I am using the term *performative* here in the sense given to it by Mieke Bal; one aspect of Bal's discussion of the performative is the concept of the theatrical *mise-en-scène,* which she redefines provisionally as "the materialization of a text – word and score – in a form accessible for public, collective reception" (97). *Mise-en-scène* is performative in Bal's terms to the extent that it provides a space or environment in which something takes place on a social scale, in a "limited and delimited section of real time and space" (97). Cage's work similarly performs weather at a formal, material level in a social space, and functions as a textual *mise-en-scène* for the operation of climatic conditions.

Leeches

"US troops were expected to patrol in towns, villages, delta regions, and paddy fields that were found throughout the South and in jungles and the bush. The climate alone could be a major drain on a soldier's physical capabilities let alone the fact that he was searching for an enemy....

"'The heat and rain and insects were almost worse than the enemy. Drenched in sweat, the men waded through flooded paddies and plantations, stopping from time to time to pick leeches out of their boots'" (Trueman).

Environmental Impact

"Geography was not the only factor that dictated the battle. The weather in the monsoon season gave the Vietnamese a slight advantage. The constant rain and low cloud ceiling prevented U.S. Close Air Support and slowed down the re-supply...

"Not only did the weather play havoc on airmobile operations but it also prevented many normal operations for soldiers in the field. On days with low cloud ceilings it could be difficult to get a good azimuth on a compass and keep track of a unit's position because of limited visibility to navigate off of surrounding terrain. In cases of severe storms lightning strikes could knock out vital communications and play havoc on a perimeter defences, because the electrical discharge from a lightning strike could set off all the claymore mines and other defence mechanisms placed outside the perimeter to prevent enemy infiltrations. Perhaps one of the most noted effects of Vietnamese weather on troops was the extreme heat and humidity that encountered the troops as the [sic] debarked the planes when arriving in the RVN. A U.S. Soldier coming fresh off the plane in Vietnam may have put it the best when he said, 'The heat just hits you right in the face. It's like California, but far more intense'" (Lydic 12).

KNOWLEDGE

Cage's appropriation of Thoreau in *Lecture on Weather* must be seen within the temporal structure that connects historical texts to the present. Cage piles up huge amounts of textual information about the weather, to the point where that information becomes so excessive and indeterminate that it problematizes the representation of weather altogether. The text departs from its historical source from the outset, because instead of following Thoreau's procedure of referencing an object in the natural world or presenting a topological scene,

Lecture on the Weather self-consciously produces knowledge about language itself – that is, the poem foregrounds the textual, linguistic character of our construction of the weather.

BOREDOM, WEATHER

"The more narcotizing the effects which cosmic forces have on a shallow and brittle personality is attested in the relation of such a person to one of the highest and most genial manifestations of these forces: the weather. Nothing is more characteristic than that precisely this most intimate and mysterious affair, the working of the weather on humans, should have become the theme of their emptiest chatter. Nothing bores the ordinary man more than the cosmos. Hence, for him, the deepest connection between weather and boredom" (Benjamin *Arcades* 101-2).

"BOREDOM + ATTENTION = BECOMING INTERESTED" (CAGE)

Another "writing through" of Thoreau, the 1975 text *Empty Words,* was intended by Cage to last 10 hours in performance, although at the infamous Naropa reading of 1975 Cage decided to limit his performance to the final, two-hour-long section of this text. He performed in a nonexpressive monotone with his back turned to the audience, and punctuated his reading with long periods of silence:

"u   tl th /   rk   /   rmn   ei"

                                        (*Empty* 68).

The audience reacted by throwing objects at the stage, playing the guitar, making bird whistles, and screaming, while Allen Ginsberg and other friendly listeners formed a circle around Cage for his protection (Silverman 266). The performance (and reaction) is not untypical, as Cage often recommended the acknowledgment of boredom as a route toward interest: "If something is boring after two minutes, try it for four. If still boring, then eight. Then sixteen. Then thirty-two. Eventually one discovers that it is not boring at all" (*Silence* 93). Here the mindful acknowledgment (and embrace) of boredom allows one to shift into a more positive mental state.

THE UNBORING BORING

Weather reports flow past us; they are always new, but always the same. They provide a limited vocabulary and a limited array of possibilities for each season, with very little scope for narrative progression. Goldsmith's *The Weather* foregrounds this endless flow of the same, this eternal return, and his text illustrates just how boring and banal the media's commodified representations of weather can be. In *The Arcades Project* file entitled "Boredom, Eternal Return," Benjamin writes that boredom is "an index to participation in the sleep of the collective" (108). The social collective is asleep, bored, drugged by the utopian promise of commodities, lulled into slumber by endless repetitions of the same: "Why does everyone share the newest thing with someone else? Presumably to triumph over the dead. This only where there is nothing really new" (*Arcades* 112). The very idea of newness, of progress, has within it the dialectical traces of regression. The question then arises: Is Goldsmith dreaming the dream of commodity culture? Is he obeying that culture's demand for novelty by representing boredom in a new way – in his own terms, by presenting the novelty of the "unboring boring" – a kind of boredom that is "fascinating, engrossing, transcendent, and downright sexy... the sort of boredom that we surrender ourselves to when, say, we go to see a piece of minimalist music" (68). Perhaps Goldsmith's reframing of the banalities of the weather within the discourse of poetry wakes us up from the dream of media commodification; to awaken from that dream, we must "surrender" to the unboring boring, and read *The Weather* (à la Benjamin) as a mass of quotations that short-circuit the eternal return of the seemingly new. If the old (weather report) is inherent in the new (weather poetry), Goldsmith's text faces two ways.

This double orientation maps out a particular ambiguity of consciousness. On one hand, the text makes us aware of the political and historical identity offered by media (the dream state of commodities, the utopian potential of the commodity fetish), while on the other hand, *The Weather* foregrounds the critical capacity to question that identity through dialectical quotation: "The new, dialectical method of doing history [of doing poetry?] presents itself as the art of experiencing the present as waking world, a world to which that dream we name the past refers in truth" (*Arcades* 389).

EVERY EPOCH

"Every epoch not only dreams the next, but while dreaming impels it towards wakefulness" (Benjamin, *Charles* 176).

———————

From: Kenneth Goldsmith [ubu@ubu.com]
Sent: 14 January 2011 14:13
To: Peter Jaeger
Subject: Re: John Cage?

It in lecture on nothing in silence

———————

From: Peter Jaeger [p.jaeger@roehampton.ac.uk]
Sent: 14 January 2011 8:48 AM
To: Kenneth Goldsmith
Subject: RE: John Cage?

Hi Kenneth,

I hope you are very well. I'm writing you because I'm trying to track down a quotation by Cage which you cite in at least two of your essays: "John Cage said, 'If something is boring after two minutes, try it for four. If still boring, then eight. Then sixteen. Then thirty-two. Eventually one discovers that it is not boring at all.'

I suspect this quotation may be in *For the Birds,* but I can't find it – you wouldn't happen to have a more exact reference, would you? Any help would be greatly appreciated

Many thanks
Peter

————

Consider the environment. Please don't print this email unless you really need to.

This email and any attachments are confidential and intended solely for the addressee and may also be privileged or exempt from disclosure under applicable law. If you are not the addressee, or have received this email in error, please notify the sender immediately, delete it from your system and do not copy, disclose or otherwise act upon any part of this email or its attachments.

FIGHT WARM-WEATHER BOREDOM WITH NEW ACTIVITIES

"School is almost over with summer approaching. Instead of sitting at home bored out of your mind, here are some things to add to your summer bucket list. These ideas are perfect if you are envying your friends who are traveling to faraway places. They are a great way to get out of the house. They also give you something to do with your friends. Why not have a paint war? All that is needed is paint, brushes and old clothes. Cups can also be used to throw the paint. Glitter bombs can also be created from real eggshells. It is the same idea; fill a clean egg with glitter. Then you smash them on somebody's head. Having a paint war is not only fun but it is a great excuse to attack your friends with paint" (Harriott).

BAD WEATHER

"Among all the subjects first marked out for lyric expression by Baudelaire, *one* can be put at the forefront: bad weather" (*Arcades* 111).

CLIMATE CHANGE DENIAL

The American historian of science Naomi Oreskes undertook a Web of Science search using the keyword phrase "global climate change" and found 928 papers. "How many of these papers present evidence that refutes the statement: 'Global climate change is occurring, and human activities are at least part of the reason why'? The answer is remarkable: none" (Oreskes 71).

PREFACE TO LECTURE ON THE WEATHER

"Our leaders are concerned with the energy crisis. They assure us they will find new sources of oil. Not only will earth's reservoir of fossil fuels soon be exhausted: their continued use continues the ruin of the environment" (Cage, *Empty* 4).

Toronto on Track for 2nd Coldest Winter in 25 Years

"If the rest of January remains below seasonal, and February is close to its average temperature, Toronto will have a mean winter temperature of –6C, making this one of the coldest winters this city has had in a quarter of a century. The average winter mean temperature is –4.1 C. The coldest winter in the past 25 years is the 1993-1994 season when the mean temperature was –7.8 C. However, since Environment Canada began keeping winter records in Toronto – in the 1937-38 season – this season may not even rank in the top 10 coldest winters. The coldest winter on record for Toronto was 1976-77, which saw a mean temperature of –8.4C. It's still cold on Wednesday, with an extreme cold alert first issued by the city on Monday still in effect" ("Toronto on Track").

Toronto Just Misses Heat Record

"July 21, 2011: Toronto likes to think of itself as the hottest spot in the country but on Thursday that title became official. With the temperature reaching 37 C and the *humidex* making it feel like 51 C – Canada's largest city became Canada's hottest city, at least for one day. Even before the sun came up Toronto had already set one record: the hottest night ever recorded in the city. The overnight low was a sweltering 26.6 C. Before that the record was 26.3, set in August 2006. Environment Canada put out a weather warning because of the extreme heat and humidity, warning people that 'during times of high heat and humidity, it is critical to stay properly hydrated by drinking plenty of fluid like water or juice.' Senior climatologist Dave

Phillips said the heat and humidity is stuck over southern Ontario and moving very slowly: 'It's very sluggish. It just doesn't move. It's like a sumo wrestler. It just kind of stands its ground and nothing can push it out of the way.' Paramedics were sent to homes, apartments and workplaces to deal with heat-related emergencies. Patricia Anderson, who is in charge of the city's cooling centres, says they've seen a steady stream of customers—almost 700 people had come to the centres by midday. 'People come in, they get a glass of water and just have a little rest,' she said" ("Toronto Just Misses").

TORONTO COOLING CENTRES

Metro Hall Cooling Centre, 55 John Street at King Street West

Centennial Park Recreation Centre, 1967 Ellesmere Road, west of Dolly Varden Boulevard

Driftwood Community Centre, 4401 Jane Street, north of Finch Avenue

East York Civic Centre, 850 Coxwell Avenue, south of O'Connor Drive

Etobicoke Olympium, 590 Rathburn Road at Melbert Road

McGregor Community Centre, 2231 Lawrence Avenue East, east of Birchmount Road

North York Civic Centre, 5100 Yonge Street, north of Sheppard Avenue West

REFRAMING SOURCES

For Kenneth Goldsmith, Cage's influence as composer, poet, and philosopher from the 1940s to the early 1990s
"cannot be underestimated" ("Why" xx). Goldsmith's 2005
book *The Weather* references Cage's *Lecture on the Weather,*
because both texts use constraint-based methods for recast-
ing preexistent source material. Yet unlike Cage's mimetic
performance of the weather – his simulated "feel of weather
in all its uncertainty and changeability" (Perloff, "Moving" 76)
– Goldsmith simply transcribes a year's worth of hourly radio
weather reports. Robert Fitterman has discussed various
modes of textual appropriation, and clarified the key difference between earlier collage forms and the approach taken
by Goldsmith and some other recent writers. Whereas the col-
lagist brings "appropriated materials together... to a singular
expression," writes Fitterman, the plagiarist presents source
material in large, unmodified chunks such as a whole paragraph or even an entire page lifted directly from a book, thereby reframing works that already exist "in new contexts to give
them new meanings" (15). Cage's appropriation of source texts
is "plagiarist" in Fitterman's sense to the extent that his work
frequently draws from single texts, such as Thoreau's *Journal*
or Joyce's *Finnegans Wake,* rather than collaging together language from a number of different texts (as might be found, for
example, in Ted Berrigan's *The Sonnets* or John Ashberry's *The
Tennis Court Oath).* However, Cage does not strictly adhere to
the conceptual plagiarist's use of unmodified textual material,
for his "writing-through" of source texts entails recycling them
in an altered form; for all his indebtedness to Duchamp and

the ready-made art object, Cage's use of chance operations modifies them beyond the scope of recontextualization. In contrast, Goldsmith lifts huge lumps of material – sometimes even entire texts, such as his appropriation of weather reports or of a single issue of the *New York Times* in *Day* (2003). For both Cage and Goldsmith, cultural production consists of the selection and reframing of source material, albeit in very different manners.

WEATHER REPORT

"A couple of breaks of sunshine over the next few hours, what little sunshine there is left. Remember this is the shortest day of the year. Looks like the clear skies hold off till later on tonight. It will be brisk and cold, low temperatures will range from twenty-nine in some suburbs to thirty-eight in midtown. Not a bad shopping day tomorrow, sunshine to start, then increasing clouds, still breezy, with a high near fifty. Couple of showers around tomorrow night, er, tomorrow evening, into early tomorrow night, otherwise partly cloudy later on, low thirty. For Monday, windy and colder with sunshine, a few clouds, high forty-two. And then for, er, Christmas Eve, mostly sunny, but with a chilly wind, high near forty degrees. For Christmas itself, cloudy with chance for rain or snow, high thirty-six. Forty-three degrees right now and cloudy, relative humidity is fifty-five percent in midtown. Regarding the current temperature forty-three going down to thirty-eight in midtown" (Goldsmith, *Weather* 3).

Context Transforms Content

"My work was only sometimes that of identifying, as Duchamp had, found objects" (Cage, *Writings*). After meeting with Marcel Duchamp in 1937, Benjamin noted his interest in a *pochoir,* or miniature replication of Duchamp's *Nu descendant un escalier.* This image consists of a black-and-white photograph of the painting, which Duchamp had altered through the application of ink and gouache. For Perloff, Benjamin's enthusiasm for Duchamp "reflects the aesthetic that governs the entire *Arcades Project*: to copy and reproduce one's own earlier words or the words of others can be a very fruitful exercise. If the 1912 Nude was 'shocking,' Duchamp asks, what about this miniature produced twenty-five years later? Has its delicately altered colour substantially changed it?... The lesson here is that context always transforms content" (*Unoriginal* 48).

Recycling

Harriet Tarlo points out that the "recycling" of found text is particularly prevalent in "experimental poetry which has a philosophical or political engagement with the environment and/or ecology" (115). In ecological terms, the recycling of found text could further be reconsidered as a strident critique of the disposable goods that our economic situation consistently produces, and which consumer culture uses without regard for environmental sustainability.

Historiography

Goldsmith's reframing of the weather report in the context of poetry represents a reconfiguration of source text in order to make apparent new possibilities for poetry. This approach parallels the logic found in Benjamin's reframing of 19th-century commodity culture—that is, his use of the "dialectical image" to illuminate the overdetermined causalities of the present moment. Historiography for Benjamin is not a matter of structuring the world as the conflict of discrete and totalized social periods. It is instead a re-production of social relations, figured among jostling textual citations. "The material of history," writes James L. Rolleston, is for Benjamin "language, the multiple interlocking languages of the past that have never ceased speaking and that fundamentally condition the historian's own language" (15). And much like Benjamin's understanding of how historical languages condition the historian's language, Goldsmith's understanding of poetic language is conditioned entirely by his sources; in *The Weather*, Goldsmith takes on the mantle of poet as historian of everyday media discourse, and he voices that history through direct quotation.

110-WS-25 Modular Weather Station

"The *110-WS-25 Modular Weather Stations* are meteorological systems designed to be a user friendly solution for data storage and real-time monitoring of weather conditions. The standard sensor package includes 6 weather parameters: Wind Speed and Direction, Temperature and Relative Humidity, Barometric Pressure, and Precipitation, (additional

optional sensors can be added), a 5-Foot Tripod and Vertical
Mast for sensor mounting, and a Data Logger with LCD Display. No computer is required for setup and viewing data. A
simple menu interface using the LCD display and three front
panel buttons makes setup easy. Data is recorded directly to
a Secure Digital (SD™) card, providing convenient data down-
loads and storage for many months of data. The data logger
is compatible with standard SD cards up to 2GB. Logging at 1
minute intervals, a 2GB card will store over 5 years of data. A
new file is created and saved to the card each day" (NovaLynx).

ECOLOGY OF THE EPHEMERAL

Benjamin's version of natural history aimed to show the
impermanent character of objects, to pull from the fragments
of commodity culture "the fleeting quality of truth that signals
the possibility of the monadic fragment against the commodity 'progression' that gave rise to it" (Mortimer-Sandilands
222). The *Arcades Project* thus provides us with "an ecology of
the ephemeral, the monadic, the truth revealed as fragment"
(222). And it is this fragmentary truth – a small T truth, a truth
of impermanence, an ephemeral truth – that characterizes the
ecology of Goldsmith's poem. What could be more ephemeral
and impermanent than a weather report?

THE WEATHER AS COMMODITY

Just as Benjamin reframes the outdated objects of nineteenth-century commodity culture, Goldsmith reframes the

disposable commodities of media information, turning the weather into a dream. Here it is necessary to jump ahead from the nineteenth-century objects of Benjamin's Paris, to consider instead the contemporary, technologically mediated objects of media information. For in our era, the weather report is a form of commodity, supported as such through paid advertising, in the case of commercial stations, or through government funding, as in the case of the BBC or public radio. The weather – like traffic and sports, two other objects of Goldsmith's reframing practice – accompanies the commercial production of desire. Like all commodities, however, the weather forecast rapidly becomes obsolete; all things are necessarily short-lived in the accelerated tempo of consumer culture. And as Goldsmith reminds us, "Nothing has less value than yesterday's news" ("Uncreativity" 1).

THE MYTHOLOGY OF WEATHER

In moving toward built-in obsolescence, commodities disclose "secret connections to the mythological, that which is ancient and out of reach" (Leslie 4). In other words, the desire for commodities (in this case, the media-based commodity of information) cannot be fulfilled; one can never stop the commodified flow of information, or possess it as object. And when Goldsmith recontextualizes the weather report as a poem, his text unmasks the public display of commodity fetishism, where wish images develop into fragments of a potential ideal: "Mostly clear, comfortable conditions tonight, and another nice day coming up tomorrow" (*Weather* 71). The poem mythologizes the weather, but in a direction that is dialectically opposed to the weather's mythical status as nature. The poem does not transform the weather into a mythical dream-object that can be consumed passively, but resituates the weather as part of a larger myth of revolutionary rupture: the revolutionary, avant-garde poem. Goldsmith's appropriative tactics paradoxically mobilize the old (nature) to make a break with the new (civilian and military information technology, mass media).

Operations Desert Shield and Desert Storm

"Weather conditions were a major concern during Operations Desert Shield and Desert Storm. One observer with experience in the region noted that coalition forces would 'face a steep learning curve in coming to grips with weather prediction throughout the area of operations.' Air Force climate data shows an average of 50% cloud cover over Baghdad in December and January, declining to almost 40% in February. This season is characterized by fog, low ceilings, clouds and rain. None of these conditions are conducive to bombing or photo reconnaissance. In April, the landscape is regularly lashed by high winds averaging 35 to 45 miles per hour that stir up savage desert sand storms. The storms are frequently followed by the *haboob*, which is the Bedouin term for 'the worst possible combination of things.' These brown thunder storms race along, embedded in the sandstorm itself, dumping tons of mud-laden water as they go. A typical storm will cover an area of 60 to 100 miles in diameter, and last from 36 to 48 hours. Under such conditions, radar is hampered and re-connaissance is difficult. However, Lt. Gen. Kelly asserted that 'We have also very closely studied the weather and the terrain, and we don't see anything there that would inhibit us from accomplishing our mission. The sandstorms are normally very localized – you can fly around them, you can work through them, you can go to a different time, so I don't really believe that would cause us a problem'" (Federation of American Scientists).

BAGHDAD WEATHER

"Military experts had warned that the attack should not be delayed until the hot season (which comes in early May in Iraq and is long and intense), and late March was already borderline. Baghdad weather bulletins, in any case, suddenly infiltrate the New York weather news, even as our troops were infiltrating Iraqi soil: 'Oh we are looking at, uh, weather, uh, across, uh Iraq obviously here for the next several days, uh, we have, uh actually some good, good weather is expected. They did have a sandstorm here earlier, uh, over the last twelve to twenty-four hours those winds have subsided and will actually continue to subside. Uh, there will be enough of a wind across the southern portion of the country that still may cause some blowing sand tomorrow. Otherwise we're looking at clear to partly cloudy skies tonight and tomorrow, uh, the weekend, uh, it is good weather and then, we could have a storm, uh, generating some strong winds, uh, for Sunday night and Monday, uh, even the possibility of a little rain in Baghdad' (39). This passage nicely exemplifies the powers of 'mere' transcription, mere copying, to produce new meanings. From the perspective of the weather forecaster, Iraq is experiencing some 'good good weather' – good visibility, no doubt, for bombing those targeted sites, and not too much wind" (Perloff, "Moving" 77).

WEATHER REPORTS IN THE GULF

"Israel and Saudi Arabia began censoring their weather condition and forecast information broadcasts shortly after the start of the war, to deny this data to the Iraqi military. Iraq was capable of receiving weather imagery transmitted by American low altitude weather satellites, and the data could be useful for Iraqi military flight or Scud missile planning and for other military purposes potentially harmful to US troops in the region. Unlike military weather satellites, civilian satellites transmit unencrypted imagery and data, making it impossible to deny this information to Iraq without also denying it to others in the region, including American military forces" (Federation of American Scientists).

READERSHIP / THINKERSHIP 1

"A certain type of book is being written that's not meant to be read as much as it's meant to be thought about" (Goldsmith, *Uncreative* 158).

DEMATERIALIZATION

Sol LeWitt's "Paragraphs on Conceptual Art" (1967) illustrates how conceptual art of the 1960s prioritized idea over
material form. LeWitt dismisses the art object as a secondary
afterthought: "When an artist uses a conceptual form of art,"
he writes, "it means that all of the planning and decisions are
made beforehand and the execution is a perfunctory affair.
The idea is a machine that makes the art" (5). Shortly after
LeWitt's article was published, Lucy Lippard coined the now
well-known phrase "the dematerialization of the art object"
(6) in order to emphasize the importance of the art idea as opposed to the documentation of that idea through language,
photography, or another medium. One significant outcome of
this dematerialization is the critique that it offers to socially
reified assumptions about value – in other words, dematerialization calls into question the cultural legitimation of art
or text as *product*, as commodity. Instead, dematerialization
foregrounds the *process* of the artwork's creation, and resituates art as research on the social and aesthetic conditions
of its own production. Almost thirty years later, Goldsmith's
rewrites LeWitt's paragraphs as "Paragraphs on Conceptual Writing" by substituting words about "writing" for words
about "art," thereby "translating" his source text into a different, albeit related, field of cultural production. For example,
whereas LeWitt writes, "When an artist uses a conceptual
form of art, it means that all of the planning and decisions are
made beforehand" (6), Goldsmith writes, "When an author
uses a conceptual form of *writing*, it means that all of the
planning and decisions are made beforehand" (98; emphasis

added). Goldsmith's translation of LeWitt is not entirely unlike Guy Debord and the Situationists' practice of *détournement* – the deflection, diversion, and appropriation of texts through modifying already existing elements. The Situationists' radical *détournement* of comic strips, advertisements, and maps during the 1950s and '60s would seem to be a significant pretext for Goldsmith's appropriation of LeWitt, with the very significant exception that Goldsmith's text implies a critique of mainstream, commercial publishing, which is predicated on such values as authenticity, individuality of voice, psychological realism, unique inspiration, and the importance of the finished, well-crafted object.

Since conceptual writing is "usually free from the dependence on the skill of the writer as craftsman" ("Paragraphs" 98), it does not fall into the ideological assumption that values craft over concept. For Goldsmith, "It is only the expectation of an emotional kick, to which one accustomed to Romantic literature is accustomed, that would deter the reader from perceiving this writing" ("Paragraphs" 98). Of course, there is one very significant difference between Goldsmith's and the Situationists' practice of *détournement.* Where the Situationists *détourned* material from popular culture, Goldsmith *détournes* a text by LeWitt that is already critical of established art-making practices. The Situationists targeted texts drawn from mass-market advertising and popular culture – texts that they regarded as supports for a social order that they were completely opposed to. Unlike the Situationists' practice, however, Goldsmith does not critically oppose his target text. Conceptual poetics is entirely sympathetic to the goals and practices of

LeWitt and the first generation of conceptual artists – that is, art should be impersonal research that should not elide process and social context.

READERSHIP / THINKERSHIP 2

While Cage's work shows some affinity to conceptualism due to its use of impersonal compositional methods, Cage took issue with conceptual art's stress on dematerialization. Daniel Charles points out that several conceptualists viewed Cage's practice favourably, because works such as his silent piece 4'33" demonstrate the notion that nothing may remain of the art object but the idea or concept (Cage, *For* 152-53). Perhaps the form of conceptualism that Charles cites holds some parallels with Goldsmith's call for a conceptual writing that is not meant to be read as much as it is meant to be thought about. Yet Cage disagrees strongly with this earlier form of dematerialization in conceptual art, because it "obliges us to imagine that we know something *before* that something has happened. That is difficult, because the experience is always different from what you thought about it" (*For* 153). For Cage, silence is not the same as the idea of silence, and thinking does not equal reading.

PROPERLY CITED

"The greatest book of uncreative writing has already been written. From 1927 to 1940, Walter Benjamin synthesized many ideas he'd been working with throughout his career into a singular work that came to be called *The Arcades Project*... It's a massive effort: most of what is in the book was not written by Benjamin, rather he simply copied texts written by others from stack [sic] of library books, with some passages spanning several pages. Yet conventions remain: each entry is properly cited, and Benjamin's own 'voice' inserts itself with brilliant gloss and commentary on what's being copied" (Goldsmith, *Uncreative* 109).

Downwind Message

"Every six hours the Air Force's chemical warfare defence installation calls for a *downwind message* which takes into account wind direction and strength" (Federation of American Scientists).

Works Cited

Bal, Mieke. *Travelling Concepts in the Humanities: A Rough Guide.* U of Toronto P, 2002.

Benjamin, Walter. *The Arcades Project.* Translated by Howard Eiland and Kevin McLaughlin, Belknap, 1999.

_____.*Charles Baudelaire: A Lyric Poet in the Era of High Capitalism.* Translated by Harry Zohn, Verso, 1983.

Cage, John. *Empty Words: Writing '73-'78.* Wesleyan UP, 1978.

_____. *For the Birds: In Conversation with Daniel Charles.* Marion Boyars, 1981.

_____. *Lecture on the Weather.* C. F. Peters, 1975.

_____. "Mureau." *M: Writings '67-'72.* Wesleyan UP, 1973.

_____. *Silence.* Wesleyan UP, 1961.

_____. *Writings through Finnegan's Wake.* Printed Editions, 1978.

Calderbank, Michael. "Surreal Dreamscapes: Walter Benjamin on the Arcades" *Papers of Surrealism 1,* https://www.research.manchester.ac.uk/portal/files/63517385/surrealism_issue_1.pdf Accessed April 24, 2014.

Federation of American Scientists. "Weather," *FAS Space Policy Project Desert Star,* http://www.fas.org/spp/military/docops/operate/ds/weather.htm. Accessed May 2, 2014.

Fitterman, Robert. *Rob the Plagiarist.* Roof Books, 2009.

Goldsmith, Kenneth. *Day.* Figures, 2003.

_____. *The Weather.* Make New, 2005.

_____. *Uncreative Writing: Managing Language in the Digital Age.* Columbia UP, 2011.

_____. "Uncreativity as a Creative Practice." *Electronic Poetry Center,* http://writing.upenn.edu/epc/authors/goldsmith/uncreativity.html. Accessed April 13, 2014.

————. "Why Conceptual Writing? Why Now?" *Against Expression: Contemporary Conceptual Poetics,* edited by Kenneth Goldsmith and Craig Dworkin, Northwestern UP, 2011, pp. xvii-xxii.

Harriott, Cassie. "Fight Warm-Weather Boredom with New Activities." *Jag Wire,* 1 May 2014, www.thejagwire.net/2014/05/01/fight-warm-weather-boredom-with-new-activities/. Accessed April 12, 2014.

Leslie, Esther. "Walter Benjamin's Arcades Project." *Militant Esthetics,* 11 May 2010, www.militantesthetix.co.uk/waltbenj/yarcades.html. Accessed May 3, 2014.

LeWitt, Sol. "Paragraphs on Conceptual Art." *Artforum* 5.10 (1967): 8.

Lewallen, Constance. *Writings Through John Cage's Music, Poetry, and Art.* Ed. David W. Bernstein and Christopher Hatch. U of Chicago P, 2001. 234-243.

Lippard, Lucy. *Six Years: The Dematerialization of the Art Object from 1966-72.* Praeger, 1973.

Lydic, Jeffrey A. "Environment and War: The Impact of Geography, Terrain, and Weather on the Vietnam War." 13 April 2008, http://webspace.ship.edu/ajdieterichward/Exhibits/Capstone/LydicExhibit.pdf. Accessed April 24, 2014.

Mortimer-Sandilands, Catriona. "Thinking Ecology in Fragments: Walter Benjamin and the Dialectics of (Seeing) Nature." *Eco Language Reader,* edited by Brenda Iijima, Nightboat Books, 2010, pp. 211-26.

NovaLynx. "110-WS-25 Modular Weather Stations." https://novalynx.com/store/pc/110-WS-25-Modular-Weather-Stations-p1073.htm. Accessed April 1, 2014.

Oreskes, Naomi. "The Scientific Consensus on Climate Change: How Do We Know We're Not Wrong?" *Climate Change: What it Means for Us, Our Children, and our Grandchildren.* Edited by Joseph F.C. DiMento and Pamela Doughman. MIT P, 2007.

Perloff, Marjorie. "Moving Information: On Kenneth Goldsmith's The Weather." *Open Letter,* vol. 12, no. 7, 2005, pp. 75-85.

_____. *Unoriginal Genius: Poetry by Other Means in the New Century.* U of Chicago P, 2000.

Rolleston, James L. "The Politics of Quotation: Walter Benjamin's Arcades Project." *PMLA,* vol. 104, no. 1, 1989, pp. 13-27.

"Seeking Sanctuary: Draft Dodgers." CBC Digital Archives. www.cbc.ca/archives/topic/seeking-sanctuary-draft-dodgers. Accessed April 13, 2014.

Silverman, Kenneth. *Begin Again: A Biography of John Cage.* Alfred A. Knopf, 2010.

Spahr, Juliana, Mark Wallace Kristen Prevallet, Pam Rehm Eds. *A Poetics of Criticism.* Leave Books, 1994.

Tarlo, Harriet. "Recycles: the Eco-Ethical Poetics of Found Text in Contemporary Poetry." *Journal of Ecocriticism,* vol. 1, no. 2, 2009, pp. 114-30.

Thoreau, Henry David. *Walden,* edited by J. Lyndon Shanley, Princeton UP, 1971.

"Toronto Just Misses Heat Record." *CBC News,* 21 July 2011, www.cbc.ca/news/canada/toronto/toronto-just-misses-heat-record-1.980629. Accessed May 21, 2014.

"Toronto on Track for 2nd Coldest Winter in 25 Years." *CityNews Toronto,* 22 January 2014, www.citynews.ca/2014/01/22/toronto-on-track-for-2nd-coldest-winter-in-25-years/. Accessed May 26, 2014.

Trueman, C. N. "War in Vietnam." *History Learning Site,* 18 December 2019, https://www.historylearningsite.co.uk/vietnam-war/war-in-vietnam/. Accessed May 24, 2014.

Process, Person Erasure: Recovering the There There in Dorn, Orange, Long Soldier & Cooperman

Matthew Cooperman

*If you free yourself from the conventional reaction to a quantity
like a million years, you free yourself a bit from the boundaries of
human time. And then in a way you do not live at all,
but in another way you live forever.*

—John McPhee, *Basin and Range*

Late afternoon, driving west in Wyoming, I-80, a vast snowy landscape unfurls, but sun streams through the window and, inexplicably, it's not windy. We're on our way (my wife, the poet Aby Kaupang) to Saratoga Hot Springs for a little R&R (sans children) after a brutal stretch of teaching and grading. We are giddy with the space, the sun, the lack of wind, an impending mineral soak. And we feel sufficiently blotted out by the landscape that the daily or casual or causal complaints of the world fall away.

It's a feeling I've experienced for over thirty years now, largely and often while driving in the West, generally from Colorado to California and back again. It's a quality of western topography so particular that it becomes archetypal. Something like *full empty.* The sheer pleasure of erasure, of landscape at speed so fast it turns space into time, is delicious, enlarging, obliterating. It bowls me with the sheer horizon of being, palpable and panoptic: "There is your domain. / It is the domicile it looks to be / or simply a retinal block of seats in." This, from the beginning of Ed Dorn's *Gunslinger,* a phenomenological game of appearance versus reality, hums in my ears, blinks in my eyes, "the enormous space / between here and formerly."

Dorn's conflationary pivot – turning space into time – is both a move apt to describe the continuum of western landscapes and a kind of necessary (and ironic) revision to Olson's dictum, stated boldly in *Call Me Ishmael,* "I take SPACE to be the central fact for man born in America man from Folsom cave to now forward. I spell it large because it comes here large. Large and without mercy" (17). Taking a sense of the projective, of Melville's peculiar projectivity in *Moby Dick,* Olson extrapolates America's geologic presence into national character. When he writes this – in 1947, against the backdrop of World War II –space feels particularly imperiled as the UR condition of experience. The atom bomb has been dropped, and the nuclear shadow shadows all.

Yet by the time Dorn is writing *Gunslinger* (1968–74), the recognition of a global holism – largely abetted by the accelerating technologies of television, film, radio, a generally conglomerated media – has emerged to the point where we are not surprised to casually see the other side of the world. "What is time but an impertinence?" Williams says in *Kora in Hell* (42). It's the age of Apollo 17's "blue marble" photo of earth or eating dinner by the blue glow of the four corners, "sick-sties" appetites experienced simultaneity as flags on coffins coming home from Vietnam (Dorn, *Gunslinger,* 103). Thus:

> Time is more fundamental than space.
> It is, indeed, the most pervasive
> of all categories
> in other words
> there's plenty of it.
> And it stretches things themselves
> until they blend into one,
> so if you've seen one thing
> you've seen them all. (5)

We exist at the same time on the same earth. Stretched, pulled, unbounded, blurred, "here and formerly" blend, on this sunny afternoon, in the suspended animation of the windshield. It's also an idea of landscape at the speed of a car, another abetting technology that make the appearance / reality game palpable, a virtual landscape that neither weathers nor starves.

But back to space-time: it's manifold, of course; we feel it. I am reminded of Olson's extraordinary formulations in "A Bibliography on America for Ed Dorn," a piece Olson wrote for Dorn *on* his impending journey west, and *for* a lifetime of study. And how a bibliography is itself a kind of time-space, what anyone's long graph of reading, rereading, annotating amounts to. Olson describes it as attention, the occasion of a life, a life dedicated to deep study. Further, that "millennia & person... are not the same as either time as history or as the individual as single." These two facts are juxtaposed, and "that henceforth one must apply to quantity as a principle... and to process as the most interesting fact of fact (the overwhelming one) how it works" (297). These four terms, *Millennia, Person, Quantity,* and *Process,* intersect in the work of art, in the subject to be discovered, in the person to be discovered: "Applying all four of these at once (which is what I mean by attention), the local loses quaintness by the test of person (how good it is for you as you have to be a work of your lifetime" (298). This has always struck me as a powerful way of capturing imagination's durational flow, its interstitial weave of subjective and objective experience, place and person gathered over time and space... a life. It's also a way of reading landscape, of reading into and beyond the local, and making one's way by landscape, a kind of dead reckoning beyond mere appearances.

Of course, there is a real landscape there, and our removal from its material forces is another way of describing the success of Manifest Destiny. Obvious and complete, closed and replete, with rest stops and warm beds, Golden Corral specials. Part of the fracture in our ecologies is the distance between our ideas about landscape and the thing itself. But Olson's axis can help us see the landscape vertically (a real achievement in an overwhelmingly horizontal world) and see through the idea of Manifest Destiny back to the landscape itself. This basin and range country is sere and sparsely vegetated, but it is hardly empty: sage grass, grama grass, salt grass, cheat grass

(invasive), desert mallow; kangaroo rats, horned lizards, taran-
tulas; black-tailed jackrabbit and desert cottontail; pronghorn
antelope, mule deer, coyote, mountain lion; western mead-
owlark, mourning dove, black-billed magpie, turkey vulture.
And cattle and sheep (invasive). And of course, Cheyenne,
Arapaho, Paiute, Shoshoni (first nation).

❈ ❈ ❈

In my book bag, nestled conveniently near the cooler and
gourmet food stuffs, is Tommy Orange's debut novel, *There
There*. Nominated for a Pulitzer, and scorching the best-seller
list, it is at once a counternarrative by and about urban Indians, a ghost dance against cultural disappearance, a portrait
of gun violence in America, and a searing rebuke to the official
narrative of American history. It's not a comfort – *There, There*
– but a comment, wryly stolen from Gertrude Stein, about the
absence of Native Americans in anything but symbols and res-
ervation clichés. There's no *there* there to Native history, no
sustaining corpus of culture in anything resembling its diverse
past.

> There was an Indian head, the head of an Indian, the drawing
> of the head of a head-dressed, long-haired Indian depicted,
> drawn by an unknown artist in 1939, broadcast until the late
> 1970s to American television everywhere after all the shows
> ran out. It's called the Indian Head test pattern. If you left the
> TV on, you'd hear a tone at 440 hertz—the tone used to tune
> instruments—surrounded by circles that looked like sights
> through riflescopes. There was what looked like a bull's eye in

the middle of the screen, with numbers like coordinates. The Indian's head was just above the bull's eye, like all you had to do was nod up in agreement to set the sights on the target. This was just a test. (3-4)

Unfortunately not a test, there's no happy ending to *There There*, just as there's no way to reverse the facts of an atrocity still abstractly acknowledged by the American populace, but there's a way to write back, to take back the story. I see that as Orange's startling achievement, validating not Native American dignity, but the contemporary facts of the legacy of conquest: double-consciousness, generational violence, alcoholism (and FAS), poverty, addiction, suicide, drug violence, gun violence. Not the story we want, perhaps, but a real story. And all in that last west place, California, and in the city, Oakland, on what was once Miwok land. Hardened to these facts, Orange says it clearly through the voice of one of the book's elders: "It's to prepare them for a world made for Native people not to live but to die in, shrink, disappear" (165).

There There approaches this tragedy ethnographically, developing the context out of which the singular violence of the novel turns. The novel starts dramaturgically, with a detailed cast of characters. Each character then speaks in first person, shifting by chapter, and there's a character, Dene Oxendene, who's collecting Native stories in a video archive. These ethnographic tropes are countervailed by a prologue and interlude, third-person insertions on Native history, naming practices, powwows, blood purity, etc. Creating a kind of thick description, these facts of conquest work on and around the ostensi-

MATTHEW COOPERMAN

ble story. The effect is both intensely personal and collective in its recursive familiarity. And the story compounds, about violence against women, the Native trap of the military, PTSD. Orange's strategy is to go right into the heart of it, like Ginsberg entering Plutonium's core, revealing the strange fruit of contemporary Manifest Destiny.

> When they first came for us with their bullets, we didn't stop moving even though the bullets moved twice as fast as the sound of our screams, and even when their heat and speed broke our skin, shattered our bones, skulls, pieced our hearts, we kept on, even when we saw the bullets send our bodies flailing through the air like flags, like the many flags and buildings that went up in place of everything we knew this land to be before. The bullets were premonitions, ghosts from dream of a hard, fast future. (10)

Importantly, Orange sources his tragic vision to the American Indian Movement and its 1970 occupation of Alcatraz. Some of the characters in *There There* are children, taken to the island by their elders and thereby intricately stitched to that Civil Rights watershed. But it is largely the aftermath of that moment into the contemporary that we get in the novel, and it reveals the enduring, intergenerational curse that is Columbus's "discovery" of North America. As Fina Gomez, one of the many grandmother figures in the book, says, "It is a real curse, and a real curse is more like a bullet fired from far off" (171).

❁ ❁ ❁

We're experiencing a new wave – the fourth – of a Native American Renaissance, and it's produced some extraordinary literature recently. I think of the powerful counternarratives of prose writers like Tommy Orange and Terese Marie Mailhot (Heart Berries), and poets like Jennifer Foerster (*Leaving Tulsa, Bright Raft in the Afterweather*), Sherwin Bitsui (*Dissolve, Shapeshift*), Joan Naviyuk Kane (*Hyperboreal, Milk Black Carbon*), and especially, Layli Long Soldier, whose 2017 National Book Critics Circle Award–winning title *Whereas* made clear that reparations are far from over. Each of these writers recalibrates the story of North America's myriad places and people. They consider quantity, the extended set, and they foreground process as the necessary duration for space to catch up with time. And in extending space – offering new places and peoples of attention – they reframe the narrative of identity: who and what and how is an American. Whereas writing. I think this work anneals, however slightly, the ruptured ecology of North America. More often than not, it is a violent story. "Now / make room in the mouth / for grassesgrassesgrasses," says Long Soldier at the beginning of *Whereas*. It's at once a reenactment of Lincoln's 1862 sentence of thirty-eight Lakota to hang for stealing food (the local merchant Andrew Myrick had said of the starving Natives, "Let them eat grass"), an ironic echo of Whitman's nation-building poem, and a clearing of the mouth for the speech of the earth.

More literally, grass functions as a metonym in *Whereas* – for resistance, or indigeneity, or prior existence, or an alternate attention. In a review on the back cover of the book, the poet Fred Moten declared, "Indigeneity is the non-locality in which we now must make a living in common. But we have no right to we on and under the ground of this history, this genocide, this language. We have no right to *now*." Reclaiming the where and the as, the place and the simultaneity, *Whereas* provokes a new language, a new commons, somewhere between we and now, here and formerly. As in:

> grass needle tips
> around the edges
> of wounds this summer
> potent
> grass songs
> a grass chorus moves shhhhh
> through half-propped
> windows I swallow
> grass scent the solstice
> makes a mind
> wind makes it
> oceanic blue a field in crests (31)

As in:

> *I don't trust nobody*
>   *but the land* I said
> I don't mean
> present company
> of course
> you understand the grasses
> hear me too always
> present the grasses
> confident grasses polite
> command to shhhhh
> shhh listen (32)

As in:

> The Dakota 38 refers to the thirty-eight Dakota men who
> were executed by hanging, under orders from President
> Abraham Lincoln.
>
> To date, this is the largest "legal" mass execution in US
> history.
>
> The hanging took place on December 26, 1862—the day
> after Christmas.
>
> This was the *same* week that President Lincoln signed the
> Emancipation Proclamation...
>
> The signing of the Emancipation Proclamation was includ-
> ed in the film *Lincoln;* the hanging of the Dakota 38 was not.
> (49)

Whereas writing, counterfactual composting, a reclama-
tion project of millennia, person, quantity and process always
already there. Paradoxically, it's a social poetics that resists
from and within landscape, a prior condition of ecological
awareness that decenters master Judeo-Christian narratives
and their systematic cultural erasures. It is also an expression
of time-space, and of the nonlocality of genocide that exists
as a consequence of a planetary singularity. The Holocaust is
with us, the Armenian genocide is with us, the Rwandan genocide is with us, the global refugee crisis is with us.

Whereas writing reveals what has not been said, acknowledged, brought into view. Just consider the roll call of atrocities precipitated on Turtle Island worth revisiting: 1492 (Co-

lumbus), 1607 (Jamestown), 1620 (Mayflower's arrival), 1675 (King Philip's War), 1754 (French and Indian War), 1830 (Indian Removal Act), 1838 (Trail of Tears), 1862 (Dakota 38), 1864 (Sand Creek Massacre), 1887 (Dawes Act), 1890 (Wounded Knee), 1906 (Antiquities Act), 1953 (Tribal Termination Act), 1975 (Pine Ridge Shootout). These dates – a partial list – are emblems of annihilation and exist in a simultaneity unresolved. They remain topological fissures in the consciousness of North America and must be reckoned by us all. As James Baldwin observed nearly seventy years ago in *Notes of a Native Son,* "People are trapped in history and history is trapped in them."

❀ ❀ ❀

A different – and perhaps surprising – kind of whereas writing occurs in Ed Dorn's *The Shoshoneans: The People of the Basin-Plateau.* After a beautiful reprint and expansion by the University of New Mexico Press, the book is now widely available. Published in 1966, replete with photographs by Leroy Smith, *The Shoshoneans* documents a journey into the Great Basin of Nevada, looking for the face and trace of the Shoshoneans. Pioneering as ethnography, it gathers Dorn's and Lucas's large personalities into a kind of road book of encounters about as different from *On the Road* as you could get. Picture a long-haired Dorn and an afroed Lucas, ostensibly on a journalistic assignment for the publisher W. W. Morrow, driving around the northern Great Basin of Nevada and Idaho, picking up hitchhikers, mostly Shoshoni, drinking beer, disappearing into remote regions of the reservations... against the building fervor of the civil rights era. The book reads meticulously as ethnography, toggling back and forth between subjective and objective points of view, noting the landscape, the morphology, the people, always building context, but never revealing an overt purpose. The purpose was in the wander, getting lost, in a landscape almost limitless in scope but connected by over five hundred years of occupation by the Shoshoni people. The trajectory of the drive is a more or less irregular circuit between the Paiute, Western Shoshoni, Shoshoni-Bannock, and Eastern Shoshoni Reservations. And while the landscape is sparsely populated and largely desert, it is deeply marked by history. What there is there in the legacy of conquest?

For Dorn, the corridor of expansion was the text to be read. Arterial pathways of game trails and rivers become Native hunting paths become trapper and explorer passage, and then the Conestoga wagon, and then the belching railroad. Or the smaller pieces that knit the material landscape together: barbed wire, windmills, irrigation ditches, the square cut nail, the steel plow. The legibility of the Basin-Plateau landscape reveals the fault lines of its ecology. That the Shoshoni reservations are on those fault lines performs the whereas story. Again and again, the book's "encounters" reveal the tensions of transcultural exchange, but do so largely without commentary. Dorn generally resists making anything symbolic, while Lucas's photographs reveal the natural symbol of Shoshoni poverty. Here's the opening passage, with Dorn and Lucas, after some rambles with a weathered Shoshoni rodeo cowboy named Wilbur, visiting the oldest living Shoshoni: "Willie

Dorsey is 102 years old. He was born before the treaty of Fort Bridger. He is a Shoshoni" (9). And:

> I begin where it was the highest pitched for me, at Duck Valley, on the Nevada-Idaho border. One hot July afternoon we stopped at Mountain City for a sandwich. It was a particularly western restaurant, varnished pine, fixed in the style of "mountain-outwest" – what the stereotype dictates but which is not really so common... We had a beer. There were Indians, I looked at them by now almost casually as my special subject, other men seemed part of the blurred background. (9)

And later, meeting Dorsey:

> He made a shaking motion with his finger, trembling in space, toward a chair. The room itself was overpowering. I was struck right off by his beauty, the sense of the power of his presence I later remember I felt immediately, but I also saw myself as a curious paleface. My attention was suddenly arrested – this man was a great deal more than old. I was looking at the scene, and at myself, in a mirror, seeing the looking. (11)

Dorn's intense self-consciousness is mixed with the thrill of the encounter, which, despite his inclinations (Edward Dorn, 1929–99, perhaps America's most iconoclastic poet), leaves him in a spell of almost religious awe. And yet:

> I was then preoccupied with what was going on, and coincident was the feeling, quite strong and uncomfortable, that I was thinking about it, again the psychological double-mirror. It was very hot in their cabin. Should we be there? There was

in me an oppressive thrill over the idea of my own presence. I
thought of it as a ruptured cord in the consciousness, a strong
confusion of the signals of my own culture. (13)

Strong confusion of the signals one's own cultural narrative: whereas writing. Resistance, iconoclasm, estrangement,
placelessness, nomadism... qualities of imagination that evoke
Ed Dorn and his writing. His and Lucas's wandering amounts
to a kind of drifter accounting in a region whose topography
has no seeming destination, no obvious there there. This, in
that methodological acuity Olson so stressed, amounts to a
kind of un-map. Matthew Hofer, in his appendix essay to the
reissue of *The Shoshoneans* (which he edited), notes:

> The condition of having no place, one Dorn frequently as-
> sumes, complicates the idea of being found just as it facilitates
> that of being lost. The value of traveling outward without nos-
> talgia – without any prospect of nostos, or homecoming – re-
> sides in a hope of encountering something "initial," something
> "without cultural reference." (98)

Dorn himself notes this wanderlust as a condition of lost-
ness in the poem "The 6th," stating "few / people are as lost
as I am," and "Everywhere I am I feel I am everywhere else"
(Derelict Air, 168). We are not too far from *Gunslinger*, which
itself dissolves before reaching its ostensible destination of
Las Vegas. Just as the greatest folly in Gunslinger is to be fixed,
described, so too in *The Shoshoneans* is any explicit purpose to
be offered. To eliminate the draw, interpretation, plan, map,

to dance with a "marvelous accidentalism," this is the highest
good. Here's the Zlinger at the end:

> Many the wanders this day I have seen
> The Zlinger addressed his friends
> Keen fitful gusts are whispering here and there
> The mesas quiver above the withdrawing sunne
>
> .   .   .   .   .   .
>
> But now niños, it is time for me to go inside
> I must catch the timetrain
> The parabolas are in sympathy
> But it grieves me in some slight way
> because this has been such fine play
> and I'll miss this marvelous accidentalism (198)

In a clear sense, the wandering in *The Shoshoneans* prefigures *Gunslinger,* but in an entirely different manner. Never sa-
tirical, *The Shoshoneans* has a moral purpose of self-examina-
tion that seems of a different racial universe. It is a pursuit of
"initial" experience outside of cultural narratives, a blankness
of reference that is also an erasure of self.

The spirit of this can be found in the Ghost Dance Movement, which originated with the nineteenth century Paiute
elder Wavoka. Essentially a vision of white decolonization, it
also stressed intertribal cooperation and sober living. Viewed
as a serious threat of insurrection, particularly after the Dawes
Act of 1887, it was ruthlessly put down at the Wounded Knee
Massacre of 1890. Were Dorn and Lucas, in tracing the origins
of Wavoka from Pyramid Lake east into the Great Basin and
across multiple reservations, restaging the resistance of the

Ghost Dance? However appropriative, was this non-Native wander open to them? Something of that whereas energy deeply permeates *The Shoshoneans.* The Acoma poet Simon Ortiz, in his foreword to the reissue, suggests as much:

> When I think about it, I have to consider that *The Shoshoneans* was also part of that voice from within the American community of that time, especially because the US Civil Rights struggle led by Dr. Martin Luther King had been waged for more than ten years then.... For me, as an Indigenous tribal community person, when I think more about it, I have to say personally it was "the community," the *hanoh* as we call it at Acoma Pueblo, that stimulated the voice and movement. And this is what inspired Ed Dorn to write *The Shoshoneans,* I believe. (6)

❀ ❀ ❀

In reading and teaching *The Shoshoneans,* in driving the arterial pathways of the West, I thought a lot about this. I thought about my time studying with Ed Dorn at the University of Colorado in the late '80s/early '90s; how "on assignment," we'd drive I-80, sniffing for the trace of the Frontier. We'd pull off at a lonely gas station, drive around back, and voilà, there'd be the wagon, the stone foundation, the hollowed log drinking trough. And of course, the Native campground, the gathering site along rivers, threads of cottonwood. I thought, too, about how the first Gulf War hovered around my time there. And how Ed was writing *Abhorrences,* a chronicle of the "long '80s," as a reaction to that time. And I thought about how that's shaped my morality, my obligations in poetry.

In 2017, I started working with Dorn's *The Shoshoneans* in a project that would be called S H O S H   N E   N S, eventually. Could I work Olson's axis of Millennia/Person/Quantity/Process? Could I continue the bibliographic life projects I'd learned from Olson and Dorn? An erasure project working off of pages of Dorn's text, the eventual chapbook stretched to thirty pages. Utilizing a process of successive "rinsing" of a key early Willie Dorsey page of the text, I attempted to get to some originary core, a preincursion core. I didn't know exactly where it would go, but let's say I wanted to get back to the grasses. Or find a way to return the page to a pre-Columbian state of population. The fact that the chap's covers are suffused with wildflower seeds and are meant to be returned to the earth – hence Plantable Press – makes terrestrial reparations possible. Erasure as strategy of delocalization, of decolonization, depopulation, of solidarity with the earth. This, sustained over time, could it perform a kind of erasure of the palimpsest of Empire? Erasing the violence, America's gun obsession. Or erasure itself, a whereas dance.

Here are two pages of the project. From this here to something ten pages later as something else.

He was   the oldest living being in
Idaho or Nevada
                along an irrigation canal
                the wheels of   the car

                the center of the valley floor—willows along

                        clouds of mosquitoes
        lava rock

        horizons
                                        Hat Peak
                        three houses, wooden clapboard
                        the first store
        the Indians had adopted that style from the settlers

                        habit of building
                                style
        resemble   community project houses

    at the edge of
                        Willie Dorsey lived in one of these
                                Lombardy poplars. The trees

                                        of great circumference
                the bowl of the valley these groves occur
                        there is a house
        cool vaulting of trees
                        three gates to be opened and then closed
                                                hooked by
                                        stick and hoop of wire.
        There were dogs   Indian
                        mongrels of
        mongrels

through the last gate to a well and short handled pump

It didn't occur to us
Willie Dorsey might live there

a hot
afternoon   the whole affair had in it the press of lateness
of last minute attention. In fact
when I did go to the screen

looked through the screen
saw   the very very old woman, the oldest creature I ever

Something ten pages later as:

lava

the bowl of

Dorsey

Working further on the project, I realized I had a beginning
with no ending. Where does the story go? What is the fate of
this journey? Performing a chance page turn, I discovered, late
in the book, the return of Willie Dorsey. How could I honor
that return? And how was this erasure a way of evacuating
Dorn's inherent racism, his treating "the Indians now almost
casually as my special subject (9)." Could I operate on my own
white privilege, my own racism, in working on Dorn? Erasure
as corrective, erasure as dispersal, erasure as initial blankness,
the salve of white space between the letters of the official account.

❀ ❀ ❀

We have nearly arrived at our destination, Saratoga,
known now for the Million Dollar Mile of fly-fishing on the
North Platte, and in the nineteenth century, as the sluice for
rail ties to the transcontinental railroad corridor. Time for a
soak, but the road stretches further, to South Pass, and then
on into the Great Salt Basin, and from there, into the Basin-Plateau, an x-axis to North American time-space if there ever was
one. Now to "free myself from conventional reactions [to]
quantity" (McPhee). I am following my own bibliography, I am
reading the grasses where I am. What follows are four pages
of the second half of *S H O S H   N E   N S.* May we start again,
whereas, in the strong confusion of the signals of our culture.

❀ ❀ ❀

Finally, these are pictures of a people who are still very potential in this hemisphere. Not Indians, certainly. That's the famous mistake Columbus made. They aren't Indians and they aren't, equally, *Americans,* another Italian assumption. It is up to them to say who and what they are. Saying and being can hardly be divided. All "American Indians" have *their own* names for themselves, naturally enough. And their names are not Shoshoni, Cheyenne, Apache, Sioux, Seminole, Caddo and what not. Those names were given to them by the various people who thought they were Indians.

I have tried to speak not of given Shoshoneans because they must be left men, women and children, in a stricter sense than It is the habit of government, agency and citizen to remember. No first names, please. No conniving familiarity. Willie Dorsey Is so intensely Shoshoni he is now a part of the world and earth, the form of his death is organic. In him we are honored to witness the total exclusion of the private. If we happen to find ourselves in the frame of a human integrity of that order, we can go openly toward it. Heaven, or nothing less. There is an apprenticeship of the spirit as well as of carpentry. All religions know that, even if the more democratic ones hide it from their novitiates. Hard work is not the answer. Of all things, diligence is hateful to the spirit. And one's senses are so in debt, an open mind is a boast too idle to do more than laugh at. But it has to be better than that. What the European found here was a collection of cosmologies he thought was a continent. He saw Indians and many tribes of those. The nuances of language articulate each spirit differently but he never knew the languages beyond the few jargon mutations needed for bargaining. Just as he never knew when he crossed the Basin-Plateau those people were anything but "poor" or, if he were eloquent, "wretched." And it is still his yardstick of Indianness – the poor Indian is more Indian, to him. The American hegemony was always economic and pseudo-real. The "successful" Indian is less Indian. And less likely to land in trouble.*

––––––––

*   The police in Pocatello have a reputation for extreme brutality toward Indians who are picked up on the street and thrown into the "drunk tank" at the city jail. I can't say for certain how much the reputation is earned. But hardly a week goes past without report of an Indian subjected to beating, always severely, sometimes fatally. The local newspaper is evidence enough that, in Traffic Court, Indians are consistently given higher fines and longer jail terms than whites for the same offenses.

                    pictures   a people   very potential

    this hemisphere

    Columbus made

                who   what
        are   being
                        their own names
                not Shoshoni, Cheyenne   Seminole

                tried to speak   Shoshonean

        men   and children
                                                    to remember

                                            Willie Dorsey

                world and earth

                                                            private

                                    ourselves
    a human

                nothing
                        the spirit as   carpentry
        religions

                answer

an open mind is

						better

cosmologies			a
	continent		saw							nuances
					differently
languages

						the Basin-Plateau
	anything
						Indian is more Indian

		American
pseudo-real

				a people

	Columbus made
			Italian

					being
		their own
	are not Shoshoni

		speak

Willie Dorsey   intensely

a human

the spirit

answer

better

cosmologies

mutations

anything

less

trouble

being

are

Dorsey

a

cosmologie

mutation

Works Cited

Baldwin, James. *Notes of a Native Son.* Beacon, 1955.

Bitsui, Sherwin. *Dissolve.* Copper Canyon, 2018.

―――. *Flood Song.* Copper Canyon, 2009.

Cooperman, Matthew. *S H O S H   N E   N S.* Grama/Plantable Press, 2019.

Dorn, Ed. *Derelict Air: From Collected Out.* Enitharmon, 2015.

―――. *Gunslinger.* 50th anniversary ed., Duke UP, 2018.

―――. *The Shoshoneans: The People of the Basin-Plateau.* U of New Mexico P, 2013.

Foerster, Jennifer. *Bright Raft in the Afterweather.* U of Arizona P, 2013.

―――. *Leaving Tulsa.* U of Arizona P, 2013.

Kane, Joan Naviyuk. *Hyperboreal.* U of Pittsburgh P, 2013.

―――. *Milk Carbon Black.* U of Pittsburgh P, 2017.

Long Soldier, Layli. *Whereas.* Graywolf, 2017.

Lopez, Barry. *The Rediscovery of North America.* Vintage, 1992.

McPhee, John. *Basin and Range.* Farrar, Straus & Giroux, 1982.

Olson, Charles. "A Bibliography on America for Ed Dorn." *Collected Prose.* U of California P, 1997.

―――. *Call Me Ishmael. Collected Prose.* U of California P, 1997.

Orange, Tommy. *There There.* Penguin Random House, 2018

Williams, William Carlos. *Kora in Hell.* Four Seas, 1920.

Virtual Water Alarms

HIROMI SUZUKI

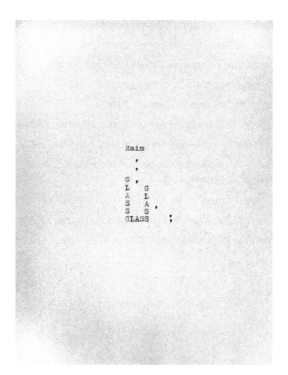

```
G
R   G
A   R G
S   A R G
S   S A R G
GROUNDWATER
        ! ! ! ! ! !
            , , ,
                ,
```

Some

Maja Jantar

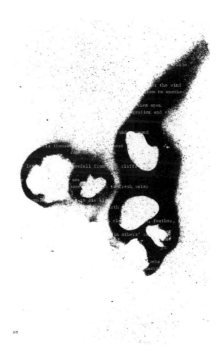

NOTES ON THE POEM

This poem was created in April 2020 and can be considered a digital collage that includes spores on paper. This piece is a collaboration with wind and spores released by the fungus *Daldinia Concentrica*, also known as King Alfred's cake, cramp balls and coal fungus. The text appears in the spore traces.

The Dioramas

ALISON HAWTHORNE DEMING

1.

The stomachs of sardines
                    are filled with prediction
          I love the   ho-hum business   revived
by two German investigators
                              who reveal   the counterfeit
          a fake Fantasia   baked tea cakes
                              ingeniously worthless
          people shouted   and wept
ingesting the cloudy formula   looking for galaxies
                    bluer that average   pulsars or charms

     looking for a star
                                   as young as Homer

2.

Most people will defend   a mammal's way of life
the naked courtship   we love so well
                   partners circle each other   for hours
              a slippery hemisphere
                                        worth fighting for
          the king himself fully erect
                        forty feet above   a treetop hotel
I prefer   sponge baths in the fountain
    coriander parasols   blown inside out by the wind
          I will spare you   the patriotic saga   of the centerfold
     whereby   average white Americans
                 strike it rich   in   just under three minutes
while eating animal products   on the veranda outside that
     pretty yellow house   they would sell   in a hot minute
                  authenticity guaranteed
                        for a night at the Hollywood Hotel

3.

Gentlemen
Enclosed is my   principal consideration
            sent in the days of the wheat harvest   gold
flowing into our mouths   the International Monetary Fund
     circling   the field of Kansas   a new kind of Midas
Please send me   one bargain paradise   the skin of the last
          parakeet   and an accurate   Tinkertoy model of
     Buddha's farewell   to the animals

If by the official date   I am not delighted   in every way
   I will have my hopes and hunches
         fully illustrated   on page 92

4.

I could end here with a   global imperative   insist that
            even earthworms   can be trained
so how about you   and aim   my sturdy machine
               along a pleasing s-shaped curve
            my hypothesis propelled   on clouds of tiny
dandelion seeds to ascend   perhaps to land
                  on the basic paradox
      of the American adventure   discovered
                              by Herman Melville
  in the once-pure seas   snakes can die of their own poison
   The best rules are   get fed   stay as clean as you can
            triple your reading speed and generate enough heat
      to breakdown   the endocrine disruptors

5.

The trigger was set   some five million years ago
         I was exposed   to the problem of sexual
      dimorphism   by my maternal grandmother
whose eyes   were filled   with European dreams
                  somehow the old-timer   advanced
         my vision   beyond domesticity

to a dignified revelry   then the trouble begins
   an etiquette manual   disappears   an anthropologist
    bribes   the peace-loving hostess   with ivory
        beads   and a bell-shaped curve
  child analysts cruise   through the restaurant
o our taste   the collector who has everything   strains the
                              heart

Notes on Heliopoetics

JOSHUA SCHUSTER

What were we doing when we untethered the earth from its sun?
—Friedrich Nietzsche

✺

there's two suns, you hear,
two,
not one –
so what?
—Paul Celan

Let's look at the sun. Its "free gift from nowhere" (to re-purpose a phrase of Hannah Arendt; Arendt 2-3) of renewable energy may be the only thing that can save us from wasting our climate. Of course, we cannot look at the sun directly, so we will need to look askance or metaphorically. Let's call this figurative looking heliopoetics. Where should we begin? Here are some possibilities: ancient religions with sun deities, paganism, Egyptology, Platonism, Neo-Platonism, medieval hermeticism, medieval church and monarchic iconography,

Gnosticism, Enlightenment rationalism, Enlightenment mysticism, romanticism, theosophy, occultism, fascism, modernist science fiction, Russian futurism, twentieth century black nationalism, the nuclear age, the space race, interstellar exploration, SETI and the search for inhabited planets, contemporary solar utopianism and permaculture, Afrofuturism, solarpunk.

Who tethered the earth to someone else's sun?

The nightmosphere

A shadowless sun

The unsun unsung

If all the Day – the singsong Sun
In rain or fog or blue –
The Sun unsung the Day in song –
For who for what for you

Major energy transitions always also involve cultural transitions. Our planet must shift to a solar and wind energy system, not only to stop global warming due to greenhouse gas emissions, but also because these energies are the most available to all and implicitly democratic and non-oligarchic renewable resources. Carbon energies and economies are not just structurally compatible with capitalism; they are the foundation for extractivist, corporate-led political economies in which just a few persons and companies that have monop-

oly access to resources can amass epochal wealth that perpet-
uates a system based on "cheap natures" and cheap labour
(Jason Moore) where exploitation, theft of people's life forces,
and externalization of environmental and social costs can be
displaced onto the wider public.

What will the world of solar energy look like? Most imagi-
nations of this energy regime envision it as primarily ecotopi-
an. However, as Imre Szeman mentions, the belief "that a shift
to wind and solar power occasions a more general expansion of social justice – [is] a completely unsubstantiated view
of how energy and social possibility are linked" (11). Szeman
announces a call for "energy justice (14)," and while all roads
seem to point to solar power, there is no guarantee that solar
will be emancipatory in broader social, political, and environmental domains. One can certainly envision a world of solar
capitalism in which all the problems of today remain awash
in a sea of conspiracy rhetoric and racial violence and endless
data exploitation fueled by the new cheap nature of the sun.

Solarity diverges into: solar communism/solar fascism;
cities of the sun/black suns; left solar radicalisms/authoritari-
an solar oppressions; shared collaborative energies/all-seeing
sovereign powers; solar excesses/solar captivities; solar com-
mons/solar privatizations; a sun that exceeds all economics,
all laws of exclusion, all logics of scarcity/the sun as yet another marketplace, yet another source of inequity and conflict; a
decolonial "solar throat slashed" (Césaire)/empires on which
the sun never sets.

Poems as solar panels, a poem on every rooftop, poems
too cheap to meter.

Here comes the unsun

A sun no one/everyone can appropriate

The auratic emanation prior to any appearance as such

For a teaspoon of sun
The emperor's army gathered
Councilors silenced debate
Villagers emptied their pockets
Cities closed the gates on the citizens
The machines began to march

On the horizon, clouds retreated
Winds tied themselves in knots
Seeds unburied themselves
The animals were chased by their ghosts
All the dogs began to bark
For a teaspoon of sun

Among our solar futures, I am interested in how poetics can help us understand interstellar messaging and how interstellar messaging can help us understand poetics. Interstellar messaging is heliopoetics. Soviet astronomer Nikolai Kardashev proposed three types of civilizations in the universe: Type I civilizations could use and control all the energy available on its planet; Type II civilizations could use and control the energy of its sun; Type III civilizations could use and control all the energy in its galaxy. Some estimate our earth will be a Type I in a century or two. Kardashev thought that we would be able to detect Type II and III civilizations with our existing technology if there are any in our galaxy. But focusing just on energy is too limiting. There would also be a Type I civilization that harnesses all the creativity of its planet, a Type II that harnesses all the creativity of its solar system, and Type III that harnesses the creativity of its galaxy. Ecopoetics and heliopoetics can get us to Type I.

> Type I – flourish with the planet's unconscious
> Type II – flourish with the sun's unconscious
> Type III – flourish with the galaxy's unconscious
>
> Type I – record all conversations with the planet
> Type II – record all conversations with the sun
> Type III – record all conversations with the galaxy
>
> Type I – make planet into art
> Type II – make sun into art
> Type III – make galaxy into art
>
> Type I – throw the planet out of the planet
> Type II – build of the sun another sun
> Type III – spin the galaxy into a new galactic spasm
>
> Type I – share the planet
> Type II – share the sun
> Type III – share the galaxy

Everything radiates

The you radiates from you

The you you radiate

Works Cited

Arendt, Hannah. *The Human Condition.* U of Chicago P, 1958.

Celan, Paul. *Breathturn into Timestead: The Collected Later Poetry.* Tr. Pierre Joris. Farrar, Straus, Giroux, 2014.

Nietzsche, Friedrich Wilhelm. *Die Frölicher Wissenschaft. Neue Ausgabe.* E. W. Fritzsch, 1887.

Szeman, Imre. *On Petrocultures: Globalization, Culture, and Energy.* West Virginia UP, 2019.

THERMOGRAPHIA:
Labwork

Adam Dickinson

On July 31, 2015, in the Iranian port city of Bandar Mahshahr, the afternoon temperature reached 46°. The humidity was 50%. These weather conditions combined to create a wet-bulb temperature of 34.6°. Wet-bulb temperature is the lowest possible measurement obtainable by evaporation alone. It is determined by wrapping the end of a thermometer in a damp cloth. At a wet-bulb temperature of 35°, the human body cannot cool itself by sweating. Even healthy people sitting in the shade will die within six hours. Climate change is expected to make these conditions more frequent and widespread.

How to imagine writing with and about heat in a warming world? What forms of writing might emerge when heat and its effects are invited into the compositional process? With the assistance of a laboratory, I underwent several heat stress trials, raising my internal body temperature by around 1.5° through active and passive heating. I exercised on a stationary bicycle for 60 minutes in a chamber set at 35°C with 65% humidity. I experienced passive hyperthermia over a duration of 5 hours using a liquid cooling garment filled with hot water. Within this suit, under an additional raincoat and a reflective blanket, I exceeded the critical wet-bulb temperature of 35°. At various intervals, during all these trials, I wrote, took cognitive tests, and measured my core temperature, skin temperature, blood pressure, CO_2 uptake, and brain blood flow among other data.

My face got red. My heartrate increased. The velocity of the blood flow through my left and right cerebral arteries decreased. The volume of my carbon dioxide uptake increased. My thoughts became feverish and anxious. I wanted it to end.

DIGITAL TYMPANIC

37.8° Internal temperature, 88 heartbeats per minute, Oxygen Consumption (Litres per minute) 0.34

Right Middle Cerebral Artery Velocity (cm per minute) 57.5; Left Middle Cerebral Artery Velocity (cm per minute) 51.2

When my body
is hot
it salivates.
I try not to raise
my arms
in hunger
because sweat stains
are territorial.
My red face
is a sunburn
streaked

with the sobriety
of antiperspirant
giving up
its heat
like pavement
in a city
of advanced
infrastructural
duress.
Offramps
raise drawbridges,
parking meters
bury their dead
in rock salts
of personal freedom.
A sunburn
is an aleatoric
artform
predicated
on prostration.
Its warmth
is truck nuts
sunward, midday
at the liquor store,
an infection
of the upper
respiratory tract
candied
like a licked nipple.
We need Purell
to keep

the messages
from shedding.
My hand
is between your legs.
Is that your heartbeat
or mine?
When the reading
is complete,
please remember
to gently
remove it.

INFERNO

*38.5° Internal temperature, 88 heartbeats per minute,
Oxygen Consumption (Litres per minute) 0.34*

*Cognitive test: Assemble all occurrences of the word "burn"
in Dante's Inferno; Working Memory Overall Reaction Time
(ms) 935.7*

Burned
descending
down into this centre
from the place
thou burnest
that misery
assails me
this burning
sentence

be as burning
follow me
feet
upon the burning
sand
margins make
not burning
recent and ancient
by flames burnt
as I should have
burned
and baked myself
and people
were burned
as fire burned
ashes wholly
behoved with burns
it irks not me
I am burning
and burned
made burned
by one who held
his son
for which I left
my body
burned
above the burning
and the headaches

ENDOTHERM

38.3° Internal temperature, 107 heartbeats per minute, Oxygen Consumption (Litres per minute) 0.26

Cognitive test: This is a test of your ability to tell what pieces can be put together to make a certain figure; Detection Task Reaction Time (ms) 598.4

It's late
in the experiment
and I am staring
into fluorescent
hindsights
my daughter
calls out to me
from the uncooked
middle
of a fever
Santa Claus
has come back
to earth
from giving birth
but his clothes
are ripped
into Arabic script
he's in a wheelchair
hypodermic
needles
ridiculing him
she emphasizes
with hoarse

tranquilizer breaths
his mouth
was open
like this
like what?
dehisce
she says
and we both
lie side-by-side
trying to centrifugal
force a revision
into that old
skunk cabbage
the idea
that you deserve
anything
at all
is regurgitated
from mouthparts
long since
broken off
in the wound
I feel her forehead
and its thought
to the touch
lifting
like a quarantine

Copia

H. L. HIX

The editor of *Fractured Ecologies* formulates as its motivating question, "How does radical experimental writing contribute to the ways we think about ecology?" In what follows, I address that question not by scholarly exposition but by ecopoetic experiment. The experiment consists in imposing two procedures on four texts.

The texts are:

1. An excerpt from ch. 1 of *An Essay on the Principle of Population* (1798), in which Malthus asserts that "the power of population is indefinitely greater than the power in the earth to produce subsistence for man."

2. An excerpt from chapter 105 of *Moby-Dick* (1851), in which Ishmael reflects on whether whaling will push whales to extinction.

3. An excerpt from chapter 3 of *On the Origin of Species* (1859), in which Darwin observes "that every organic being naturally increases at so high a rate, that if not destroyed,

the earth would soon be covered by the progeny of a single pair."

4. part of a press conference (29 March 2001) by the then U.S. president George W. Bush, in which he calls for "an active exploration program... to make sure that we've got enough gas to be able to help reduce greenhouse emissions"

The texts are presented here in chronological order, and under titles drawn from them. The two procedures are these:

1. Following poet Karla Kelsey's premise that line breaks "require the reader to consider the relation of part to whole, the connection between line and sentence and stanza" (139), I impose lineation on the texts, each of which is originally presented in prose.

2. Following the example of Alice Notley's book-length poem *The Descent of Alette,* I apply quotation marks as a poetic technology toward the end, as Notley formulates it, of "measuring" the poem, making the reader "slow down and silently articulate – not slur over mentally – the phrases."

The procedures are the only alterations to the texts; I have not, for instance, changed word order or made abridgements within the passages.

By applying to these four texts those two procedures, this experiment tests the hypothesis that, applied to language that bears on ecological matters, poetic techniques aimed at achieving considered relation of part to whole and measured articulation may also help ecological reflection be more "considered" and "measured."

By calling this experiment "Copia," I mean to mark the etymological connection between *copy* and *copious,* their common derivation from the Latin *copia,* meaning "abundance" or "plenty," and thus to pose the question what relationships might hold between copying and copiousness, between citational literary techniques and the issue of abundance or its lack, an issue that haunts ecological reflection and shows itself plainly in assertion or denial of the ecological position named cornucopianism, which regards resource scarcity as an idle threat, one that human ingenuity will always avert.

Rather than imposing on the reader, as scholarly exposition aims to do, a QED, this experiment invites the reader's (re)consideration of, and (re)orientation in regard to, abundance, positioning the reader as participant in rather than recipient of the inquiry. In addition, the experiment operates as an open rather than a shut case. It purports to be not exhaustive but generative: in principle it might be applied to many texts, not only to these four, and it might be extended to include many investigative procedures in addition to these two.

These Two Unequal Powers

"I think" "I may fairly make" "two postulata."
"First, That food" "is necessary to" "the existence of man."
"Secondly, That" "the passion between the sexes"
"is necessary" "and will remain" "nearly in its" "present state."

"These two laws, ever since" "we have had any knowledge"
"of mankind," "appear to have been" "fixed laws" "of our
nature;" "and, as we have not" "hitherto seen" "any alteration" "in them, we have" "no right to conclude" "that they
will ever" "cease to be"

"what they now are," "without an immediate" "act of power"
"in that Being" "who first arranged" "the system" "of the universe;" "and for the advantage" "of his creatures," "still exe-
cutes," "according to fixed laws," "all its various" "operations."

"I do not know" "that any writer" "has supposed that" "on this
earth man" "will ultimately be" "able to live" "without food."
"But Mr. Godwin has" "conjectured that the passion"
"between the sexes may" "in time" "be extinguished." "As,
however,"

"he calls this" "part of his work" "a deviation" "into the land
of conjecture," "I will not dwell" "longer upon it" "at present"
"than to say that the best" "arguments for the perfectibility"
"of man, are drawn" "from a contemplation"

"of the great progress" "that he has already made"
"from the savage state," "and the difficulty" "of saying

where" "he is to stop." "But towards the extinction" "of the
passion" "between the sexes," "no progress whatever" "has
hitherto been made."

"It appears to exist" "in as much force" "at present" "as it did
two thousand" "or four thousand years ago."
"There are individual" "exceptions now" "as there always
have been." "But, as these exceptions" "do not appear"
"to increase

in number," "It would surely be a very" "unphilosophical"
"mode of arguing," "to infer merely" "from the existence"
"of an exception," "that the exception would," "in time,
become" "the rule, and the rule" "the exception."

"Assuming then," "my postulata" "as granted, I say," "that the
power of" "population" "is indefinitely greater" "than the
power" "in the earth to produce" "subsistence for man."

"Population," "when unchecked, increases" "in a geometrical
ratio." "Subsistence increases only in" "an arithmetical ratio."
"A slight acquaintance" "with numbers will shew" "the
immensity of" "the first power" "in comparison of" "the
second."

"By that law of our" "nature" "which makes food necessary"
"to the life of man," "the effects of these" "two unequal
powers" "must be kept equal."

"This implies a strong" "and constantly operating" "check
on population" "from the difficulty" "of subsistence."

"This difficulty must fall" "somewhere;" "and must necessarily" "be severely felt" "by a large portion" "of mankind."

"Through the animal" "and vegetable kingdoms," "nature has scattered" "the seeds of life" "abroad with the most profuse" and liberal hand." "She has been" "comparatively" "sparing in the room" "and the nourishment" "necessary to rear them."

"The germs of existence" "contained in this" "spot of earth," "with ample food," "and ample room" "to expand in, would fill" "millions of worlds" "in the course" "of a few thousand years." "Necessity, that" "imperious" "all pervading" "law of nature,"

"restrains them within" "the prescribed bounds." "The race of plants," "and the race of animals" "shrink under" "this great restrictive law." "And the race of man" "cannot, by any" "efforts of reason," "escape from it." "Among plants and animals" "its effects

are waste of seed," "sickness," "and premature death." "Among mankind," "misery and vice." "The former, misery, is" "an absolutely necessary" "consequence of it." "Vice is a highly" "probable consequence," "and we therefore see it" "abundantly prevail;"

"but it ought not," "perhaps, to be called" "an absolutely necessary" "consequence." "The ordeal of virtue" "is to resist all" "temptation to evil."

"This natural inequality" "of the two powers" "of population,"
"and of production" "in the earth," "and that great law of our"
"nature" "which must constantly keep" "their effects equal,"
"form the great difficulty" "that to me

appears" "insurmountable" "in the way to" "the perfectibility"
"of society." "All other arguments are" "of slight and subordi-
nate" "consideration in comparison" "of this." "I see no way
by which" "man can escape"

"from the weight of this law" "which pervades" "all animated
nature." "No fancied equality," "no agrarian regulations"
"in their utmost extent," "could remove" "the pressure of it"
"even for a single century." "And it appears," "therefore,"

"to be decisive against" "the possible existence"
"of a society," "all the members of which," "should live
in ease," "happiness," "and comparative leisure;"
"and feel no anxiety" "about providing" "the means

of subsistence" "for themselves and families."

"Consequently," "if the premises are just," "the argument
is conclusive" "against the perfectibility" "of the mass of
mankind."

Wondrous Extermination, Different Nature

"But still" "another inquiry" "remains; one often" "agitated
by the more" "recondite Nantucketers." "Whether owing"
"to the almost" "omniscient look-outs" "at the mast-heads
of" "the whale-ships," "now penetrating" "even through"
"Behring's straits,"

"and into the remotest" "secret drawers" "and lockers of the
world;" "and the thousand" "harpoons and lances" "darted
along" "all continental coasts;" "the moot point" "is, whether Leviathan" "can long endure" "so wide a chase," "and so
remorseless" "a havoc;"

"whether he must not" "at last" "be exterminated" "from the
waters," "and the last" "whale, like the last" "man, smoke"
"his last pipe, and then" "himself evaporate" "in the final
puff."

"Comparing the humped" "herds of whales" "with the
humped herds" "of buffalo," "which, not forty" "years ago,
overspread" "by tens of thousands" "the prairies" "of Illinois
and Missouri," "and shook their" "iron manes and scowled"
"with their thunder-" "clotted brows"

"upon the sites" "of populous" "river-capitals, where now"
"the polite broker" "sells you" "land at a dollar an inch;"
"in such a comparison" "an irresistible"
"argument would seem" "furnished, to show" "that

the hunted whale" "cannot now escape" "speedy extinction."
"But you must" "look at this matter" "in every light." "Though
so short" "a period" "ago – not a good" "life-time –" "the
census of" "the buffalo in Illinois" "exceeded the census" "of
men now in London," "and though" "at the present day"

"not one horn" "or hoof of them remains" "in all that region,"
"and though the cause of this" "wondrous extermination"
"was the spear of man;" "yet the far different nature" "of the
whale-hunt" "peremptorily" "forbids so inglorious" "an end

to" "the Leviathan." "Forty men in one ship" "hunting the
Sperm Whale" "for forty-eight months" "think" "they have
done extremely well," "and thank God," "if at last they" "carry
home the oil" "of forty fish." "Whereas, in the days" "of the old
Canadian" "and Indian

hunters" "and trappers of the West," "when the far west"
"(in whose sunset" "suns still rise)" "was a wilderness" "and
a virgin," "the same number of" "moccasined men," "for the
same" "number of months," "mounted on horse" "instead of
sailing" "in ships, would have slain"

"not forty, but" "forty thousand and more" "buffaloes;"
"a fact that," "if need were, could be" "statistically stated."

"Nor, considered aright," "does it seem" "any argument"
"in favor of" "the gradual extinction" "of the Sperm Whale,"
"for example, that" "in former years" "(the latter part"
"of the last century, say)" "these Leviathans," "in small pods,"

"were encountered" "much oftener" "than at present," "and, in consequence," "the voyages were not" "so prolonged," "and were also much more" "remunerative." "Because," "as has been elsewhere" "noticed, those whales," "influenced by"

"some views to safety," "now swim the seas" "in immense caravans," "so that to a large" "degree the scattered" "solitaries," "yokes, and pods, and schools" "of other days" "are now" "aggregated into vast" "but widely separated," "unfrequent armies."

"That is all." "And equally fallacious" "seems the conceit," "that because the so-called" "whale-bone whales" "no longer haunt" "many grounds" "in former years abounding" "with them, hence that species" "also is declining." "For they are only"

"being driven" "from promontory to cape;" "and if one coast" "is no longer enlivened" "with their jets, then," "be sure, some other" "and remoter strand has been" "very recently startled" "by the unfamiliar" "spectacle."

"Furthermore:" "concerning these last" "mentioned Leviathans," "they have two firm" "fortresses, which, in all" "human probability," "will for ever remain" "impregnable." "And as upon" "the invasion of" "their valleys, the frosty Swiss" "have retreated

to" "their mountains; so," "hunted from the savannas" "and glades of the middle seas," "the whale-bone whales" "can at last" "resort to their Polar" "citadels, and diving" "under

the ultimate" "glassy barriers and walls" "there, come up"
"among icy" "fields and floes;" "and in a charmed circle"
"of everlasting" "December, bid" "defiance to all pursuit"
"from man."

"But as perhaps fifty" "of these whale-bone whales" "are
harpooned for one" "cachalot, some philosophers" "of the
forecastle" "have concluded that" "this positive havoc" "has
already" "very seriously diminished" "their battalions." "But
though

for some time past" "a number of these whales," "not less
than 13,000" "have been annually" "slain" "on the nor' west
coast" "by the Americans alone;" "yet there are" "considerations" "which render even this" "circumstance of little" "or
no account"

"as an opposing" "argument" "in this matter."

"Natural" "as it is to be somewhat" "incredulous"
"concerning the populousness" "of the more enormous"
"creatures of the globe," "yet what shall we say" "to Harto,
the historian" "of Goa, when he tells us" "that at one hunting"

"the King of Siam" "took" "4000 elephants;" "that in those
regions" "elephants are numerous as" "droves" "of cattle"
"in the temperate climes." "And there seems" "no reason
to doubt" "that if these elephants, which" "have now been
hunted" "for thousands of years,"

"by Semiramis," "by Porus," "by Hannibal," "and by all
the successive" "monarchs of the East–" "if they still survive"
"there in great numbers," "much more" "may the great whale
outlast" "all hunting," "since he has a pasture" "to expatiate
in," "which is precisely"

"twice as large" "as all Asia," "both Americas," "Europe
and Africa," "New Holland, and all" "the Isles of the sea" "combined."

"Moreover: we are" "to consider, that from the" "presumed"
"great longevity of whales," "their probably" "attaining the
age" "of a century and more," "therefore at any one" "period
of time," "several distinct" "adult generations" "must be contemporary."

"And what that is," "we may soon gain" "some idea of,"
"by imagining" "all the grave-yards," "cemeteries,"
"and family vaults" "of creation yielding up" "the live bodies"
"of all the men," "women, and children" "who were alive"

"seventy-five years ago;" "and adding" "this countless host"
"to the present" "human population" "of the globe."

"Wherefore, for all" "these things, we account" "the whale
immortal" "in his species," "however perishable" "in his individuality." "He swam the seas" "before the continents" "broke
water;" "he once swam over" "the site of the Tuileries," "and
Windsor Castle,"

"and the Kremlin." "In Noah's flood, he despised" "Noah's
Ark;" "and if ever the world" "is to be again" "flooded,"
"like the Netherlands," "to kill off its rats, then" "the eternal
whale" "will still survive," "and rearing" "upon the topmost
crest"

"of the equatorial" "flood, spout his" "frothed defiance" "to
the skies."

We Forget That the Birds

"We will now discuss" "in a little more detail" "the struggle"
"for existence." "In my future work" "this subject shall be
treated," "as it well deserves," "at much greater length." "The
elder De Candolle" "and Lyell have largely" "and philosophi-
cally" "shown

that all" "organic beings" "are exposed to severe" "competition." "In regard to plants," "no one has treated" "this subject
with" "more spirit and ability" "than W. Herbert," "Dean of
Manchester," "evidently" "the result of his great" "horticultural knowledge."

"Nothing is easier" "than to admit in words" "the truth"
"of the universal" "struggle for life," "or more difficult"
"– at least I have found it so –" "than constantly to bear"
"this conclusion in mind." "Yet unless it be" "thoroughly

engrained" "in the mind," "I am convinced" "that the whole"
"economy of nature," "with every fact on" "distribution,"
"rarity," "abundance," "extinction," "and variation," "will be
dimly seen" "or quite misunderstood." "We behold" "the face
of nature"

"bright with gladness," "we often see" "superabundance of
food;" "we do not see," "or we forget," "that the birds which
are" "idly singing round us" "mostly live" "on insects or
seeds," "and are thus constantly" "destroying life;" "or we
forget how largely"

"these songsters," "or their eggs," "or their nestlings" "are
destroyed" "by birds and beasts of prey;" "we do not always
bear" "in mind," "that though food may be" "now superabun-
dant," "it is not so" "at all seasons" "of each recurring year."

"I should premise" "that I use the term" "Struggle" "for
Existence" "in a large" "and metaphorical sense," "including
dependence" "of one being on another," "and including"
"(which is more important)" "not only the life" "of the
individual,"

"but success in leaving" "progeny." "Two canine animals"
"in a time of dearth," "may be truly said" "to struggle
with each other" "which shall get food" "and live." "But a
plant" "on the edge of a desert" "is said to struggle" "for life"

"against the drought," "though more properly" "it should be
said" "to be dependent" "on the moisture." "A plant which

annually" "produces a thousand seeds," "of which on an average" "only one" "comes to maturity," "may be more truly said"

"to struggle with the plants" "of the same and other kinds" "which already clothe" "the ground." "The mistletoe is" "dependent on the apple" "and a few other trees," "but can only In" "a far-fetched sense" "be said to struggle" "with these trees," "for if too many"

"of these parasites grow" "on the same tree," "it will languish and die." "But several" "seedling mistletoes," "growing close together" "on the same branch," "may more truly be said" "to struggle" "with each other." "As the mistletoe" "is disseminated by birds," "its existence"

"depends on birds;" "and it may metaphorically" "be said to struggle" "with other fruit-bearing plants," "in order to tempt birds" "to devour" "and thus disseminate" "its seeds" "rather than those" "of other plants." "In these several senses," "which pass into each other," "I use"

"for convenience sake" "the general term of" "struggle for existence."

"A struggle for existence" "inevitably follows" "from the high rate at which" "all organic beings" "tend to increase." "Every being, which" "during its natural lifetime" "produces several eggs" "or seeds," "must suffer destruction" "during some period" "of its life," "and during some season" "or occasional year," "otherwise," "on the principle" "of geometrical" "increase, its numbers" "would quickly become so" "inordinately" "great that no country"

"could support the product." "Hence, as more" "individuals are produced"

"than can possibly" "survive," "there must" "in every case" "be a struggle" "for existence, either" "one individual" "with another" "of the same species," "or with the individuals" "of distinct species," "or with the physical" "conditions of life."

"It is the doctrine" "of Malthus applied" "with manifold force" "to the whole animal" "and vegetable kingdoms;" "for in this case" "there can be no" "artificial" "increase of food," "and no prudential" "restraint from marriage." "Although some species"

"may be now increasing," "more or less rapidly," "in numbers," "all cannot do so," "for the world" "would not hold them."

"There is no exception" "to the rule that every" "organic being" "naturally" "increases" "at so high a rate, that" "if not destroyed," "the earth" "would soon be covered" "by the progeny" "of a single pair." "Even slow-breeding man" "has doubled" "in twenty-five years,"

"and at this rate," "in a few thousand years," "there would literally" "not be standing room" "for his progeny." "Linnaeus"

"has calculated that" "if an annual plant" "produced only two
seeds" "– and there is no plant" "so unproductive as this –"
"and their seedlings" "next year

produced two," "and so on, then" "in twenty years there
would be" "a million plants." "The elephant is reckoned" "to
be the slowest breeder" "of all known animals," "and I have
taken" "some pains to estimate" "its probable minimum"
"rate

of natural Increase:" "it will be under" "the mark to assume"
"that it breeds" "when thirty years old," "and goes on breeding" "till ninety years old," "bringing forth" "three pairs of
young" "in this interval;" "if this be so," "at the end" "of the
fifth century" "there would be alive"

"fifteen million elephants," "descended from" "the first pair."

"But we have better" "evidence" "on this subject than"
"mere theoretical" "calculations," "namely, the numerous"
"recorded cases of" "the astonishingly" "rapid increase"
"of various animals" "in a state of nature," "when circumstances"

"have been favourable" "to them during two" "or three
following" "seasons." "Still more striking" "is the evidence
from" "our domestic animals" "of many kinds" "which have
run wild" "in several parts" "of the world:" "if the statements"
"of the rate

of increase" "of slow-breeding cattle" "and horses" "in South
America," "and latterly" "in Australia," "had not been well"
"authenticated," "they would have been" "quite incredible."
"So it is with plants:" "cases could be given"

"of introduced plants" "which have become" "common"
"throughout whole islands" "in a period" "of less than ten
years." "Several of the plants" "now most numerous" "over
the wide plains" "of La Plata, clothing" "square leagues of
surface"

"almost to the exclusion" "of all other plants," "have been
introduced" "from Europe;" "and there are plants" "which
now range" "in India, as I hear" "from Dr Falconer," "from
Cape Comorin" "to the Himalaya," "which have been import-
ed" "from America"

"since its discovery." "In such cases," "and endless instances"
"could be given," "no one supposes" "that the fertility"
"of these animals" "or plants" "has been suddenly"
"and temporarily" "increased in any" "sensible degree."

"The obvious" "explanation is that" "the conditions of life"
"have been very" "favourable," "and that there has" "consequently been less" "destruction" "of the old and young," "and
that nearly" "all the young" "have been enabled" "to breed."
"In such cases"

"the geometrical" "ratio of increase," "the result of which"
"never fails to be" "surprising," "simply explains"
"the extraordinarily" "rapid increase" "and wide diffusion"

"of naturalised productions" "in their new homes."

"In a state of nature" "almost every plant" "produces seed,"
"and amongst animals" "there are very few" "which do not"
"annually pair." "Hence we may" "confidently" "assert,
that all plants" "and animals" "are tending to increase"

"at a geometrical" "ratio, – that all would" "most rapidly
stock" "every station" "in which they could" "any how exist, –"
"and that the geometrical" "tendency to increase"
"must be checked by" "destruction" "at some period of life."

"Our familiarity" "with the larger domestic" "animals
tends," "I think," "to mislead us:" "we see no great destruction" "falling on them," "and we forget that thousands" "are
annually slaughtered" "for food, and that" "in a state of
nature"

"an equal number would have" "somehow" "to be disposed of."

"The only difference" "between organisms" "which annually
produce" "eggs or seeds by the thousand," "and those which
produce" "extremely few," "is, that" "the slow-breeders"
"would require" "a few more years" "to people," "under
favourable conditions,"

"a whole district," "let it be ever so large." "The condor" "lays
a couple of eggs" "and the ostrich a score," "and yet" "in the
same country" "the condor may be" "the more numerous"
"of the two:" "the Fulmar petrel" "lays but one egg,"

"yet it is believed" "to be the most numerous" "bird in the
world." "One fly deposits" "hundreds of eggs," "and another,"
"like the hippobosca," "a single one;" "but this difference"
"does not determine" "how many individuals"
"of the two species" "can be supported" "in a district." "A large
number of eggs" "is of some importance" "to those species,"
"which depend" "on a rapidly fluctuating" "amount of food,"
"for it allows them" "rapidly" "to increase in number."

"But the real importance" "of a large number" "of eggs or
seeds" "is to make up for" "much destruction" "at some
period of life;" "this period" "in the great" "majority of cases"
"is an early one." "If an animal can" "in any way

protect" "its own eggs or young," "a small number" "may be
produced," "and yet" "the average stock" "be fully kept up;"
"but if many eggs" "or young are destroyed," "many must be
produced," "or the species will become" "extinct." "It would
suffice"

"to keep up the full number" "of a tree, which lived" "on an
average for" "a thousand years," "if a single seed" "were produced once" "in a thousand years," "supposing that this seed"
"were never destroyed," "and could be ensured"

"to germinate" "in a fitting place." "So that in all cases,"
"the average number" "of any animal" "or plant depends"
"only indirectly on" "the number" "of its eggs or seeds."

"In looking at Nature," "it is most necessary"
"to keep" "the foregoing considerations" "always in mind—"

"never to forget" "that every single" "organic being"
"around us may be said" "to be striving" "to the utmost"

"to increase in numbers;" "that each lives by" "a struggle"
"at some period" "of its life;" "that heavy destruction"
"inevitably falls" "either on the young or old," "during
each generation" "or at recurrent intervals." "Lighten

any check," "mitigate the destruction" "ever so little,"
"and the number" "of the species will" "almost instanta-
neously" "increase to any amount." "The face of Nature" "may
be compared" "to a yielding surface," "with ten thousand
sharp wedges" "packed

close together" "and driven inwards" "by incessant blows,"
"sometimes" "one wedge being struck," "and then another"
"with greater force."

Decisions on Science, Common-Sense Decisions

"Mr. President," "in the last few weeks you have" "rolled back"
"health and safety" "and environmental measures" "proposed by the last" "administration," "and other previous"
"administrations." "This has been" "widely interpreted"

"as a payback time" "to your corporate donors." "Are they
more important" "than the American people's" "health
and safety?" "And what else" "do you plan to repeal?"

"Well, Helen," "I told people" "pretty plainly" "that I
was going to" "review" "all the last-minute decisions"
"that my predecessor" "had made, and that" "is exactly"
"what we're doing." "I presume you're referring" "to the
decision"

"on arsenic" "in water. First of all," "there had been no
change" "in the arsenic—" "accepted arsenic" "level in water"
"since the '40s." "And at the very last" "minute, my predecessor" "made a decision," "and we pulled back" "his decision"
"so that we

can make" "a decision" "based upon sound science and
what's" "realistic."

"There will be a reduction" "in the acceptable amount"
"of arsenic per billion" "after the review" "in the EPA."

"How about" "stopping the black lung benefits" "for
families?" "This is sort of—" "to increase" "some of the benefits" "of these miners?"

"We will work with" "members of the delegation" "and
make sure people" "are properly treated." "Ours is going to
be" "an administration" "that makes decisions" "on science,"
"what's realistic," "common-sense decisions."

"For example," "circumstances" "have changed since the
campaign." "We're now in an" "energy crisis." "And that's why
I decided" "to not have mandatory" "caps on CO2," "because"

"in order to meet" "those caps, our nation" "would have had
to have had"

"a lot of natural gas" "immediately" "flow into the system,"
"which is impossible." "We don't have" "the infrastructure"
"able to move" "natural gas."

"We need to have" "an active" "exploration program."
"One of the big" "debates that's taking place" "in the
Congress," "or will take place" "in the Congress," "is whether
or not" "we should be exploring" "for natural gas" "in Alaska,

for example, in" "ANWR." "I strongly think" "we should" "in
order to make sure" "that we've got enough gas" "to be able
to help" "reduce greenhouse emissions" "in the country."
"See, gas is clean," "and yet there is not" "enough of it."

"And we've got pipeline" "capacity problems" "in the
country." "We have an" "energy shortage."

"I look forward" "to explaining this today" "to the leader
of Germany" "as to why I made" "the decision I made."
"We'll be working with" "Germany;" "we'll be working with"
"our allies" "to reduce greenhouse gases." "But I will not
accept" "a plan

that will harm" "our economy" "and hurt" "American
workers."

Works Cited

Bush, George W. Exchange with journalist Helen Thomas at press conference. 29 March 2001. www.whitehouse.gov. Accessed 23 February 2014.

Darwin, Charles. *On the Origin of Species,* edited by William Bynum, Penguin, 2009.

Kelsey, Karla. "Lineation in the Land of the New Sentence." *A Broken
Thing: Poets on the Line,* edited by Emily Rosko and Anton Vander
Zee, U of Iowa P, 2011, 138-41.

Malthus, Thomas. *An Essay on the Principle of Population,* edited by Geof-
frey Gilbert, Oxford UP, 1999.

Melville, Herman. *Moby-Dick,* edited by Hershel Parker and Harrison
Hayford, W. W. Norton, 2002.

Notley, Alice. *The Descent of Alette.* Penguin, 1996.

Channels

FRANCES PRESLEY

on boulder beach

always the eye is drawn to   Barry power station

      wherever we are on this coast

                 she draws a line to   Llantwit

suck and spew of the waves   gentle brown backwash

        in the heat

robin scolds   is this summer   or winter

      white fishing boat   midway

drawn across   or pulled apart

that is the only real difference   in all this   difference

first there was a sea   then there was no sea
                        then there were trees

then there was   no leaves and seeds

            sea worms   hole   the wood
(re)peat

I must learn to see   touch and hold

            at low tide   when their world
  reaches ours

❋

anemone memory

in the rock pool   small red sea anemone   beadlet

      after fifty years it grows as big as a tennis ball

            eats dead crab meat very slowly

Ted Hughes said   *England is a sea anemone*   *which swallows*
                                                    us whole

            his hand opened   closed

see
an
enemy

It's a choice, isn't it? I said

he was red   would swallow me whole

❀

Pink Bay
                    for RM

white veins marble

                    pink rocks

        quartz veins to carry me
                                    through

                            or back to White Ladder

curvy white lines

                    not   stone
                    rows   not
                    man   made

follow this

enter this
flow

deep rock pool

guter   cwter
gutter

gryn   cryn
grin

inlet

laver mess
of seaweed potage
Margaret made

this tarp   this warp
of dark green iron flopping into pools

emerge to the white curve

querk luft ertz

cross vein ore

ice crystal
locks
in these rocks

passerine
> *for DE*

creak of gorse   cloud coming in

sheep shelter & shit   passerines hide in the gorse

two passers-by search for a path to Anthorn over mudflats

we'll phone our friends
> *they're on the other side of the country –*

decide about a smart phone   decide not

                    drink on the go   don't think on the go

rhombic antennae   Anthorn masts transmit to sub –

marine   callsign   G.   Q.   D.   – –   O

sore roar of jet plane   distant   no threat   just yet

herd not heard   the full story held off –   shore

quotes furbish furbelows   down below

                    bellow   our hollows and horrors

kid's play   kids splay   five finger exercise full fingers five   to fathom

you send   a rain damaged   past – oral   post – card

    more than mock
           do the words bounce off the hay

skein of wool   pulling us back
        skein of geese

curlew call   faint & distant   to who we are

somewhere located below   the Sol – way   cold solar
              plexus

Grune Point March 18

            ❀

Needle's Eye

line of salmon fishing stakes
        right angles
  abandoned
     no haaf netting
             across the Solway

    Needle's Eye
       arches over

incisions in the greywacke   tell tale totem
leaf motif   tweak tympanum

inca divisions   titular trouble
lost maps   twiddles his thumbs

relished device   behind a tailor dummy
tender sandstone   in tubular suit

rummaged ears   hunched collar overshadows
liquorice white spill   head lower than

quartz waterfall   testicular shrinkage

storm through the arch
up from the south

come on lass   *good girl*

   Scottish deerhound   she says

         words lost   I touch her bony skull

II

                   eye

     rise high

              fold on fold

              anticlinal
igneous

             eye rise
                   high

       pink strata

                             ❋

Notes on the Poems

These poems are part of a sequence that concerns channels of water, shorelines, and parallel coasts. The main sites are the Bristol Channel, Solway Firth, the Wash, and the English Channel, which, for nature conservation purposes, form a single entity, but politically can be sharply divided. I'm interested in how they are being shaped and formed, both by climate change and by political disruption of various kinds in the UK, especially post-Brexit. Most of the sequence was written on site and is a response to the coast, making use of visual design, verbal and typographical parallels, and slippages. I'm also working collaboratively across borders, especially in the Bristol Channel, with poets in Somerset and Wales.

on boulder beach. From the coastal town of Minehead In north Somerset, the main landmark on the opposite Welsh coast is Barry power station, but for a precise reference point, Tilla Brading drew a line on the map to Llantwit Major. A rock-pool walk with Nigel Phillips uncovered all kinds of sea creatures: oarwrack and dulce red seaweed; the dog whelk, which shelter hermit crabs; and the red sea anemone. He talked about the ancient history of the Bristol Channel when it alternated between swampy woodland and sea. As we slithered around the rockpools chewing sea lettuce, behind us Butlin's holiday camp blared out 'Dark Side of the Moon'. I think we were on the dark side of the moon, as far as they were concerned. In *Somerset's Coast: A Living Landscape,* Phillips emphasises how tightly our beaches and coastal areas are constricted by the built environment, so the sea can no longer form dunes farther back as sea levels rise as a result of polar melt. It was a warm day in November when I wrote 'on boulder beach'.

anemone memory. After I remembered this disturbing encounter with Ted Hughes, I looked for other appearances of the sea anemone in poetry and discovered some do use the anemone as a dangerous force, such as Gregory Corso's 'Active Night', in which it is a predator representing the impossibility of community between species. Thinking of Donna Haraway's critter kin, try Edwardian Edward Dowden, who you feel actually spent time with a sea anemone and calls it 'my little brother'. I also reread HD's *Notes on Thought and Vision,* where she writes about the 'overmind' state of artistic consciousness as 'a closed sea-plant, jelly fish or anemone' (19). Her analogy does not create any division between mind and body—her centres of consciousness are both the brain and the womb— nor is it antagonistic to other creatures.

Pink Bay. Pink Bay is at Porthcawl, near Swansea, on the Welsh coast, where I have collaborated with poet Robert Minhinnick. He works in Welsh as well as English language poetry, and I have used some Welsh words, such as *cwter,* inlet or channel, and *cryn,* large or strong; as well as the Saxon word for quartz. Pink Bay is at the farthest point of the Bristol Channel as the estuarine sediment begins to give way to the open Atlantic. Opposite, I could see the Exmoor hills above Glenthorne but only as dark shapes. The geology of Pink Bay creates a marbling effect of white quartz veins in pink sandstone, and it reminded me of the quartz Neolithic stone row on Exmoor, called 'White Ladder' (Presley 16-17).

passerine and **Needle's Eye.** The Solway Firth (Scottish Gaelic: *Tràchd Romhra*) forms part of the border between England and Scotland, between Cumbria and Dumfries and Galloway. Needle's Eye is a natural arch on the Dumfries Coast. Passerine was written at Grune Point in Cumbria, and it's for the poet Dan Eltringham.

Works Cited

Brading, Tilla and Frances Presley. *Stone Settings*. Odyssey Books, 2010.

Eltringham, Dan. *Cairn Almanac*. Hesterglock, 2017.

HD (Hilda Doolittle). *Notes on Thought and Vision; and, the Wise Sappho*. Peter Owen, 1988.

Phillips, Nigel. *Somerset's Coast: A Living Landscape*. Natural Time Out, 2011.

Presley, Frances. *Lines of Sight*. Shearsman, 2009.

S kin of Older Woman
Walking From

BRENDA HILLMAN

a lake   shaking   after swimming   with *if* and *is*
venules in   her seven-layered   s kin   stratum cells   her
Lamellar   stratum corneum   stratum lucidum   stacked up layers
stratum granulosum   walking from   a lake at risk   purple platelets   body's
mini-cities of troy   trees of consciousness   bent from chill   lake less clear   than
a Wednesday ago   water warmer   than decades before   post-cancer site
somewhat caved in

s kin of white woman   dripping with cold   medulla cortex   sheaths   hairs bent
bulblike   past eternal present   human   s kin   miracle   to be known   lake less clear
than decades ago   oil glands   hair roots   plunged   granular   cells of keratin
dropped as she swam   2 million-year-old   lake unplanned   doing eternity or never
on its own   nearby towns   subcutaneous   motor oil urban run-off   still swimmable water
warmer than planned   wider than Tuesday   older than   joy   older white s-kin
so oddly set   water   seems invincible   lake   too warmer   by one degree   a woman swam
her s kin loves   science   stratum spinosum   s   kin has wild nerves   making vowels   in poems
came cancer   decades   don't know when

stratum basale   algae growth   in the lake   lost thin s kin   melanocytes   light pigment
history's mistake   dermis   Merkel disks   Meissner's corpuscles   recollection when   touched by
bandaids borrowed makeup   radiant sex   said to her lover   sweet pressure   mm   nice
lake   absorbs scaly   Kokanee salmon   brought in   a decade ago   small shrimp
species   further declined   EPA spent 47 million   some s kin sloughed off
inclined blood   chagrinned   upside down   tubes of   collagen   corporations sell
embarrassment   old s-kin   different surfaces   aren't embarrassment
lake with more future   since 2010

blood vessels   5 yards of them   4 yards of nerves 650 sweat glands   curled   cobra tubes
at work   her nervousness   god forgot   to turn the science off   sweat sheaths   hair follicles
skinny tulip bulbs   under surface   1500 sense receptors   secret   pathways
with   yet yet   yet yet   thinking of
100 oil glands   nerves   touching air   after swimming   good it
feels something   a word is   the s kin of experience   spinous cells   sweat glands
hypodermis   good type of fat   papillary   capillary   Ruffini's endings   all that
walking from the lake   goose-bumps   real goose nearby   s kin rises   near friends
no longer rises for   the national anthem   rises   for   love rises with
picnic   with moonrise toward   the rest of them

for S & G

ACKNOWLEDGMENTS AND PERMISSIONS

Unless noted otherwise, images are utilized under fair use standards for purposes of criticism, education, research, and scholarship. "Dr. James P. Campbell's Safe Arsenic Complexion Wafers" from the National Museum of American History, circa 1890: https://americanhistory.si.edu/collections/search/object/nmah_1339217. "E$$o" courtesy of Greenpeace (2007): https://wayback.archive-it.org/9650/20200406054534/http://p3-raw.greenpeace.org/international/en/news/features/exxon-still-funding-climate-ch/. Scan from *An Incomplete Natural History* by Maggie O'Sullivan (1984) adapted from http://eclipsearchive.org/projects/INCOMPLETE/html/contents.html. Image of Maggie O'Sullivan's artwork *An Order of Mammal* adapted from cover of *In the House of the Shaman* (1993), with consent from Reality Street. Scans of brief fragments taken from Susan Howe's *That This* (2010). "The elastic surfaces of molecules, light" and "the film partially develops, leaving the pastoral open to fancy" taken by Orchid Tierney. Arthur Sze, "Lichen Song" and "Salt Song," from *Sight Lines.* Copyright © 2019 by Arthur Sze. Reprinted with permission. "full:new" courtesy of Jess Allen and Bronwyn Preece and made for *Fractured Ecologies.* "Virtual Water Alarms" taken by Hiromi Suzuki for *Fractured Ecologies.* "Some" produced by Maja Jantar for *Fractured Ecologies.*